Power-System Reliability Calculations

MONOGRAPHS IN MODERN ELECTRICAL TECHNOLOGY

Alexander Kusko, Series Editor

1. Solid-State DC Motor Drives by Alexander Kusko
2. Electric Discharge Lamps by John F. Waymouth
3. High-Voltage Measurement Techniques by Adolf J. Schwab
4. The Theory and Design of Cycloconverters by William McMurray
5. Computer-Aided Design of Electric Machinery by Cyril G. Veinott
6. Power-System Reliability Calculations by Roy Billinton, Robert J. Ringlee, and Allen J. Wood

Power-System Reliability Calculations

Roy Billinton,
Robert J. Ringlee, and
Allen J. Wood

The MIT Press
Cambridge, Massachusetts, and London, England

Copyright © 1973 by
The Massachusetts Institute of Technology

All rights reserved. No part of this book may be reproduced in any form or by any means, electronic or mechanical, including photocopying, recording, or by any information storage and retrieval system, without permission in writing from the publisher.

This book was set in Times New Roman
by Wolf Composition Co., Inc.,
printed by The Murray Printing Company
and bound by Dave Dunn & Company
in the United States of America.

Third printing, 1980
Second printing, 1978
Library of Congress Cataloging in Publication Data

Billinton, Roy.
   Power-system reliability calculations.

  (Monographs in modern electrical technology)
  Includes bibliographical references.
  1. Electric power systems—Reliability. 2. Electric power distribution—Tables, calculations, etc.
  I. Ringlee, Robert J., joint author. II. Wood, Allen J., joint author. III. Title.
TK1005.B57   621.319   73-2681
ISBN 0-262-02098-X

CONTENTS

Preface   vii

1 INTRODUCTION TO POWER-SYSTEM RELIABILITY CALCULATIONS   1

1.1 Introduction   1
1.2 Background and Historical Development   1
1.3 Résumé of the Contents of the Monograph   5
References   6

2 RELIABILITY AND AVAILABILITY APPLICATIONS TO DISTRIBUTION SYSTEMS   8

2.1 Introduction   8
2.2 The Renewal Process (Repairable System) Recognizing the Run-Fail-Repair-Run Cycle   10
2.3 Models for Redundant Component Overlapping Outages   20
2.4 Reliability Procedures for Substations   31
References   44

3 APPLICATION TO GENERATION PLANNING   46

3.1 Introduction   46
3.2 Generation System Model   47
3.3 Load Model   67
3.4 Expansion Analysis   79
3.5 Load Statistics   90
3.6 Interpretation of the LOLP and Margin Frequency Indices   96
References   99

4 APPLICATIONS TO BULK POWER-SUPPLY SYSTEMS   101

4.1 Introduction   101
4.2 Two-Area Reliability Evaluation   102
4.3 Composite System Reliability Evaluation   104
4.4 Conditional Probability Approach   109
4.5 Data Requirements   114
4.6 Transmission Planning   124
References   125

5 GENERATION-SYSTEM OPERATION 127

5.1 Introduction 127
5.2 Derated State Concepts 131
5.3 The Effect of Peaking Equipment on the Operating Reserve 141
5.4 Interconnected System Spinning-Reserve Requirements 158
References 169

Index 171

PREFACE

The authors of this monograph have not attempted to develop a textbook in power-system reliability calculations: It is more like a progress report, and it is hoped that the publication of this monograph will contribute to the growth of the use of analytical reliability techniques to design and analyze power systems. There are more unsolved than solved problems in this field.

Our purpose is to describe practical methods that have been utilized by the authors and their associates in solving actual power-system reliability problems. This experience has extended over a number of years and has included both work with graduate-level students in various programs and consulting engineering assignments with electric utilities in both North and South America.

The field is not new. Reliability methods have been used in the design of power-generation systems for some 40 years. The general problem is that of designing or evaluating the reliability of an ongoing system. This is not the same problem as the mission-oriented reliability problem of the space vehicle or defense electronics system that has received so much attention in recent years.

The monograph starts with applications to subsystem design by considering substation layout and generation-system reliability. These areas have received considerable attention, and the methods presented have evolved over a number of years. In the last two chapters the intent is more to give a progress report on the reliability aspects of combined generation- and transmission-system design and the reliable operation of power generation systems. There are, of course, other important reliability problems in the power system that have not been treated here.

It is our firm belief that the successful application of these power-system reliability calculation methods depends upon the simultaneous existence of three factors:

1. the development of an appropriate engineering model for the reliability problem at hand,

2. the establishment of an appropriate risk index, or index of service quality, and

3. the access of system and component operating and failure information from which reliability and availability parameters may be estimated.

The first is perhaps the easiest to accomplish. The second always involves a degree of subjectiveness that may lead to an endless debate in any specific case. The significance of the third is most often underestimated. In the past, the collection and processing of data has often lagged behind the development of reliability models. It is our hope that this monograph may in some small way serve to spur these efforts. We have attempted to include an adequate description of the required data along with the development of each model.

Finally, the authors would like to express their appreciation for the many contributions of their current and past associates. This must include those many members of groups sponsored by the IEEE, the EEI, and the CEA. Much of the material in this monograph has been taken from technical papers published by the IEEE. We wish to express our appreciation for the permission to use this material. We would also like to thank Sarah N. Ringlee for drafting the illustrations and Mrs. Roy Billinton for her assistance in the preparation of the original manuscript.

Power-System Reliability Calculations

1 INTRODUCTION TO POWER-SYSTEM RELIABILITY CALCULATIONS

1.1 Introduction

The primary objective of this text is to collect and illustrate techniques that have been applied to the prediction of reliability and availability of specific segments of an electric power system. The term reliability is used in an engineering sense to indicate a quality of constancy of service. The text attempts to emphasize the engineering and numerical procedures employed in making reliability and availability predictions. References are offered to direct the reader to fundamental developments of basic statistical and probability theory.

The terms reliability, availability, adequacy, dependability, and security are defined as needed and as applied in each specific section. The application of these terms has evolved over many decades, and therefore usage of some of the terms is unique to the area of power system applications. It seems appropriate to start this text with excerpts from notable early papers on the subject. It is possible to obtain a good idea of the models and indices employed three decades ago from comments by Dean in 1938,[1] Lyman in 1933,[2] and Smith in 1934.[3]

1.2 Background and Historical Development

Observe the use of such words as "reliability" and "effectiveness" in the 1938 paper by Dean:[1]

One of the really difficult problems faced by those responsible for planning of electric supply systems is that of deciding how far they are justified in increasing the investment on their properties to improve service *reliability*. While this problem is not at all new in the industry, it has nevertheless taken on greatly increased significance in the past few years.

In general, there are three broad aspects of this *reliability question*.

The first is to know thoroughly the *present quality* of one's service and just who is harmed by the present outages, and how much. With such a background of system performance it is not difficult to determine where in general the greater hazards lie.

The second aspect is a knowledge of the methods at hand to improve service in the many situations which arise, as well as the cost of these remedies. It is highly important that these methods of improved reliability be studied out in advance and their *effectiveness* and *cost* clearly defined. It is very easy to lapse into vague generalizations which may not, in fact, be anything like as effective as they first appear.

The third and most important is the exercise of judgement as to where and when, all things considered, expenditures should be made for *increased reliability* and how far to go with them. In theory, the criterion is that of customer's complaints and what increased price he is willing to pay for more reliability. In practice, it is very difficult to determine how many complaints concerning a given kind of trouble are sufficient to justify added expenditure to correct it, and there is no way of asking the customer how much he is willing to pay for improved reliability.

W. J. Lyman, in his prize-winning paper on power-system planning,[2] stated that

three of the most vital problems around which the whole fabric of future planning is woven are forecasting, the relation between load and capacity and fixed capital replacements.

He also stated:

A major problem in the design of a power system arises from a combination of the desire to render reasonably continuous service and the inherent fallibility of equipment. A rather large proportion of the fixed capital is so occupied, and a careful analysis of the relation between load and capacity is the starting point in an effort to reduce cost of service.

Lyman, followed by Smith[3] and Benner,[4] introduced the concepts of randomly occurring events and applied probability analysis to study the requirements for spare generating capacity. Lyman and Smith identified two classes of problems: The first is concerned with the "chance coincidence" of unrelated events, such as the overlapping, random, independent outages of a number of generators. The second problem concerns

widespread and unpredictable catastrophic events which may disable an entire generating station or even the entire system. . . . In such emergencies, the mere multiplicity of generating units or even generating stations may be of little or no avail in avoiding loss of load.[3]

Smith claimed that for the first class of problems, probability theory has its most useful application and that data for calculations for the catastrophic class were "difficult to determine."

Lyman and Smith also introduced two criteria for appraising the reliability of generation supply. Lyman studied the "probable interval between capacity outages." He reasoned that

There is very little question about providing for breakdown of one unit (boiler-turbine-generator) because this is known to occur quite often. Furthermore, reserve is usually installed for a double outage because experience has shown that this may occur every two or three years. However, very little money is spent in anticipation of a combination of breakdowns that may occur on the average of, say, once every twenty or thirty years.

Smith, on the other hand, studied the risk of losing a part of the load:

> The problem of how much spare capacity to provide resolves itself into two distinct parts:
> First, how reliable shall the service be? What expectation of load outage in a year shall be deemed satisfactory? . . . Secondly, once this standard has been agreed upon, the system should be engineered to meet it. From the coal pile to the customer's meter exists a series of apparatus, a kind of chain, each link of which may at times fail. The sum of outage expectations of each of these links must be made equal to the outage expectation set up for the system as a whole.

Although considerable attention was directed in the papers by Lyman, Smith, and Benner to the generating-capacity problem, lack of data and limitation of computational facilities severely restricted the numerical application of reliability procedures to the study of generating-system adequacy. A generating system with adequate capability is ready to serve load as necessary considering the variability of load and the variability of operational capacity depending on maintenance requirements and on unscheduled outage. It appears that probability methods were first applied to the study of spare generating capacity and that Lyman and Smith received the credit for the first proposal to utilize such methods. From his studies of the relationships between overlapping capacity outages, Smith concluded:

> It is, or at least should be well recognized, either intuitively or through actual experience, that as the number of generating units in a system increases with growth in load, or due to interconnection with other systems, the percentage of spare capacity can be decreased without sacrifice of service reliability.

These are root concepts contained in two widely used planning indices for generating systems: "interval between outages" (which necessitate curtailment of load) and "loss-of-load probability" (that is, the probability that generating capacity will be deficient).[5] In both cases, attention is focused upon events in which there is insufficient capacity available to meet the demand due to overlapping outages of a portion of the units in the generating system. The generation and loads are assumed to be connected to the same bus (single area) or, at most, a small number of buses (multiarea). The indices can reflect interarea tie-line capability, reliability, and availability but because of the single-bus assumption cannot properly recognize intra-area lines.

Practical methods for developing these indices are available.[5-8] The methods account for scheduled maintenance and overhaul requirements, annual distributions of daily peak demands, seasonal equipment loading and overloading limitations, overlapping forced-outage events, and risks of deviations in demand forecasts from realized demand.

The two methods treat independent generation-outage events, Smith's "type one" problems, and do not treat the "widespread and unpredictable catastrophic events," the "type two" problem. It is evident that Smith was concerned with problems of the type involving generation, transmission, and major substations near load centers. In a more modern context these are "bulk power supply" problems.

An attendant problem associated with the utilization of statistical concepts is the availability of applicable and consistent data, and in this regard performance records of generating units have been kept for many years.[8,9] Information suitable for generation-reserve planning, such as operating data and scheduled and forced-outage data, has been collected and published by industry organizations such as the EEI[9] and the IEEE.[10]

Three vital problems in the future planning of generation plant are the following:
1. long-range forecasting,
2. capital requirement prediction for additions and replacement of generating plant, and
3. assessment of risk of generating-capacity deficiency.

Significant steps forward in the use of probability methods occurred with the models developed by Calabrese[5] and by Halperin and Adler.[7] In both instances it appears to the authors that the key contribution was the development of a practical model and a practical index of, or measure of, generating-system adequacy. The essential element in both approaches was the separation of the generating and transmission systems, and both the Calabrese and the Halperin–Adler models concern the generating capability only. That is, they assumed the generating capability to be connected to one equivalent bus at which the load was also connected. Their assessment then was on the adequacy of generating capacity to meet load under the assumption of an adequate transmission system at all times.

The model was extended to study the import-export capability between two regions by Cook and his co-workers,[11] and a model was suggested by Szendy for a multiple-area system.[12] These models used only simple capacity criteria and linear distribution factors for power flow between the various areas and, hence, were extensions of the capacity model studies. Transmission or bulk power-supply modeling must reflect the necessity of maintaining system voltage and maintaining loadings within the thermal limits of individual circuits and components and system stability limits.[13] The bulk power-supply models must involve both static and dynamic checks, that is, load-flow evaluations with static contingencies and dynamic analysis of the system's ability to recover from specific conditions. Only limited application of quantitiative probabilistic methods has been made with bulk power-supply evaluations.[14,15] The costs of carrying out comprehensive evaluations and the lack of suitable data appear to be serious obstacles. The application of probability methods to distribution-system design extend over a period nearly as long as the applications to generation. Dean, in his 1938 paper,[1] cited studies of means for improving the frequency and duration of subtransmission and feeder outages and suggested certain goals for these parameters and means of achieving the improvement.

1.3 Résumé of the Contents of the Monograph

The purpose of this monograph is to describe methods currently applied to the assessment of power-system reliability. The monograph opens with a discussion of reliability and availability applications to transmission and distribution systems treating independent component outages and their effects upon the continuity of supply. This chapter serves as an introduction to models used in subsequent chapters and employs examples where only simple continuity or simple overload criteria are sufficient to analyze radial or simple series-parallel systems. Discussion of analytical methods of transmission systems analysis is reserved for a later chapter on the bulk power-supply problem.

The discussion in the third chapter covers models for generation planning. Capacity planning models place all generation and loads on essentially one or two buses. The fourth chapter extends the work into the area of bulk power-supply-system reliability evaluation, and methods are offered for prediction of composite reliability of the generation and transmission systems. In this case a system failure is charged if the power supply at any major transmission bus does not meet standards of voltage or thermal load on equipment. The final chapter extends the work into the operating reliability assessment concerned with generation operating-reserve problems. The major distinction between the approaches listed in Chapter 3 and Chapter 5 concerns the recognition of the current bulk power-supply state and its influence upon assessment of adequacy of the generating system a short period ahead to meet forecasted loads.

References

1. S. M. Dean, "Considerations Involved in Making System Investments for Improved Service Reliability," *EEI Bulletin*, vol. 6, 1938, pp. 491–496.

2. W. J. Lyman, "Fundamental Consideration in Preparing a Master System Plan," *Electrical World*, vol. 101, June 17, 1933, pp. 788–792.

3. S. A. Smith, Jr., "Space Capacity Fixed by Probabilities of Outage," *Electrical World*, vol. 103, February 10, 1934, pp. 222–225.

4. P. E. Benner, "The Use of the Theory of Probability to Determine Spare Capacity," *General Electrical Review*, Schenectady, N.Y., vol. 37, no. 7, 1934, pp. 345–348.

5. AIEE Committee Report, "Application of Probability Methods to Generating Capacity Problems," *AIEE Transactions*, vol. 80, pt. III, 1961, pp. 1165–1177.

6. G. Calabrese, "Determination of Reserve-Capacity by the Probability Method," *AIEE Transactions*, vol. 69, pt. II, 1950, pp. 1163–1185.

7. H. Halperin and H. Adler, "Determination of Reserve-Generating Capability," *AIEE Transactions*, vol. 77, pt. III, 1958, pp. 530–544.

8. IEEE Committee Report, "Proposed Definitions of Terms for Reporting and Analyzing Outages of Generating Equipment," *IEEE, Transactions on Power Apparatus and Systems*, vol. 85, 1966, pp. 390–393.

9. Edison Electric Institute, "Report on Equipment Availability for the Seven Year Period 1960–66," EEI Publication 67-23.

10. AIEE Joint Subcommittee Report, "Forced Outage Rates of High Pressure Steam Turbines and Boilers," *AIEE Transactions*, vol. 76, pt. III, 1967, pp. 338–343.

11. V. M. Cook, C. D. Galloway, M. J. Steinberg, and A. J. Wood, "Determination of Reserve Requirements of Two Interconnected Systems," *IEEE, Transactions on Power Apparatus and Systems*, vol. 82, 1963, pp. 18–33.

References

12. C. Szendy, "Economical Tie-Line Capacity for an Interconnected System," *IEEE, Transactions on Power Apparatus and Systems*, vol. 83, 1964, pp. 721–726.

13. R. Billinton, "Composite System Reliability Evaluation," *IEEE, Transactions on Power Apparatus and Systems*, vol. 88, 1969, pp. 276–280.

14. R. Billinton and M. P. Bhavaraju, "Transmission Planning Using a Reliability Criterion—Part I—A Reliability Criterion," *IEEE, Transactions on Power Apparatus and Systems*, vol. 89, 1970, pp. 28–34.

15. R. Billinton and M. P. Bhavaraju, "Transmission Planning Using a Reliability Criterion—Part II—Transmission Planning," *IEEE, Transactions on Power Apparatus and Systems*, vol. 90, 1971, pp. 70–78.

2 RELIABILITY AND AVAILABILITY APPLICATIONS TO DISTRIBUTION SYSTEMS

2.1 Introduction

This chapter serves as an introduction to the application and analysis of simple models based upon statistically independent repairable components. The chapter begins with a consideration of some arithmetical methods for representing the component–operation-failure-repair cycle. The models are applied successively to a simple distribution feeder, to a ring-bus transmission substation, and to distribution substations. The concepts of probability, duration, and frequency of an event are considered, and simple techniques are described for predicting the probability, duration, and frequency of events involving several independent components. The models are then extended to represent certain dependent effects such as scheduled outage of redundant components, common environmental effects that raise the risks of outage during certain periods of high stress on all redundant components, risks of overload-induced outages in redundant components, and, finally, systematic nonrandom overlapping failures of redundant components and common-mode failures. A failure modes and effects analysis is offered for a ring-bus substation as well as for industrial substation and feeder arrangements.

The indices proposed for analysis of these various arrangements are the probability of loss of service continuity, expressed by the long-term average unavailability of service; the frequency with which service continuity is curtailed; and the average duration of service continuity curtailment, given that it has occurred. A method is proposed to measure the continuity of service to a specific node or bus. The measure of continuity or its complement (the loss of continuity of service) is given by the percentage of time that the service is available (unavailable) as well as by the frequency with which service curtailment occurs and the average duration of service curtailment.

A study of the ways in which equipment and systems fail is essential to any undertaking of a reliability prediction or a reliability analysis. Fowler offers the following observations regarding failures that occurred in the space industry.[1] The position is taken that "the possibility that

Introduction

failures arise randomly, that is, without understandable cause, is excluded, but the stochastic element in failure observation is accepted." In his analysis of failures occurring in the space industry, Fowler suggested three categories within which to put the system failures:

Type 1: The system failed because it could not have worked in the first place. The major subdivisions of this type observed in practice are as follows:
a. The design is inherently incapable of performing the actual mission either because there is an unworkable combination of parts or because the system's functional logic does not correspond with the requirement.
b. The use environment was beyond the capability of the system either because it was never qualified for the actual environment or because the environment was mis-estimated.

Type 2: The system equipment could have worked if it had been just like the drawing, but it was not and, hence, failed. There are two major subdivisions which are as follows:
a. A faulty piece/part was built into the hardware.
b. The hardware was damaged in manufacture, tests, repair, or handling.

Type 3: The system could work and did work but has now worn out. The principle subdivisions of this class are as follows:
a. Some part of the hardware returned far enough toward thermodynamic equilibrium so that the hardware no longer operates.
b. Some part accumulated environmental damage to the point where it no longer performs its function.

In addition to Fowler's categories, the most important aspect of failure analysis concerns the conditions under which the failure was discovered. Green[2] discusses the aspect of whether or not a fault is revealed or unrevealed. For example, a closed breaker in an inoperative condition may continue to function quite satisfactorily until it is called upon to trip. The cause of a stuck breaker may or may not be detected depending on the tests, maintenance, and operating procedures employed.

The effect of a component outage upon a system may be quite different depending upon the nature of the arising cause. This can be seen by considering the failure modes for circuit-interrupting or circuit-breaking equipment. Two categories of breaker outages have proved useful in systems analysis. The first category involves cases where other protective equipment is required to remove the defective or inoperative breaker. For instance, if the breaker has a fault, or if it fails to interrupt, or if it fails to trip, then the resulting fault must be cleared by backup equipment. Such action must increase the extent of effect of the fault. The other class of removal corresponds to a maintenance outage or a trip-out in which the device is removed by switching and in which the extent of the outage is confined to the path involving the breaker. Events such as tests, scheduled maintenance, and scheduled repairs would be included in this category, as would false trip incidents.

Recognition must be given to those incidents that are of a temporary nature and those events that persist. Contrast the effect of a line flashover that may be cleared by tripping the circuit and reclosing as soon as deionization has taken place with those events in which permanent damage resulted, necessitating removal for repairs. Recognition must also be given to events that may affect several components in a systematic, nonrandom fashion. The term common-mode failure has been applied to these events.[3] Examples of this type of failure are readily found in the electric power industry. Capacitor failures in control equipment and solid state equipment failures due to ambient temperatures beyond equipment capabilities are two contemporary examples. The possibility of performing systematic design reviews[4] automatically follows from a consideration of this type of failure analysis.

2.2 The Renewal Process (Repairable System) Recognizing the Run-Fail-Repair-Run Cycle

A utility system is a "continuing operation," not a "mission-oriented" or "batch-process" type operation. Preventive maintenance is performed to keep equipment and lines in good condition and the failure rate correspondingly low. Repair is done as promptly as is feasible to restore service and return the system to its original capability and configuration. The power supply to customers is often backed up by additional generation and lines on alternate feeder circuits such that loss of a generator or a bulk power-supply circuit may be detected only with sensitive voltage or frequency recording equipment. In contrast, the loss of a feeder may result in an interruption of service for the duration of the time to switch to an alternate feeder or to carry out feeder repairs.

The first step in the description of the "reliability" and "availability" of a continuing type of operation is to provide a model of the run-fail-repair-run process. For purposes of analysis, the model selected is the "renewal process" in which the *stochastic processes* of failure and repair are the *same* for each failure-repair cycle.

The Renewal Process (Repairable System)

2.2.1 The Failure-Repair Cycle, Cycle Time, Availability

Suppose a line is observed for an interval of time in which N cycles of failure (permanent faults) and repair are noted. Let the observed time-to-failure for the first cycle be m_1 and the time-to-repair observed for that first failure be r_1. Similarly, let m_i and r_i be the observed times-to-failure and repair for the "ith" cycle. To defend the claim that the run-repair cycle is a "renewal process," the run-repair cycles must be statistically independent and the distribution of durations stationary in time. It is also necessary that the expected values of m and r exist such that they may be reasonably estimated by

$$\overline{m} = \frac{1}{N} \sum_1^N m_i \tag{2.1}$$

and

$$\overline{r} = \frac{1}{N} \sum_1^N r_i. \tag{2.2}$$

Note that the effects of the faults on the system have not been defined. At this point, the question is asked, Was this line permanently faulted or not? As discussed later in this chapter, a permanent fault will lead to a line outage but not necessarily to a service interruption.

The average cycle of the failure-repair process (see Figure 2.1), given by the sum of the average time-to-failure and the average time-to-repair, becomes

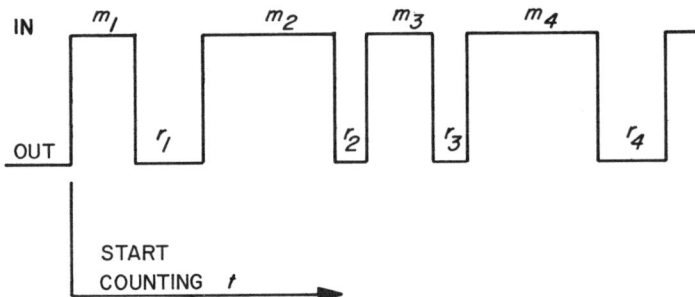

Figure 2.1 Component history.

$$\bar{T} = \bar{m} + \bar{r}. \tag{2.3}$$

The fraction of the time the line is "in," or available for service, is given by the ratio of average uptime $\bar{m}$ to average cycle time $\bar{T}$ and is termed the availability:

$$A = \frac{\bar{m}}{\bar{T}} = \frac{\bar{m}}{\bar{m} + \bar{r}}. \tag{2.4}$$

The complement, or unavailability, $\bar{A} = 1 - A$, is given by the ratio of the average downtime $\bar{r}$ to the average cycle time $\bar{T}$:

$$\bar{A} = 1 - A = \frac{\bar{r}}{\bar{T}} = \frac{\bar{r}}{\bar{m} + \bar{r}}. \tag{2.5}$$

The fault frequency f is the reciprocal of the average cycle time:

$$f = \frac{1}{\bar{m} + \bar{r}}. \tag{2.6}$$

This is the result for the "two-state model" illustrated in Figure 2.2.

2.2.2 Repairable Components in Series

Suppose two *statistically* independent devices are required to be in

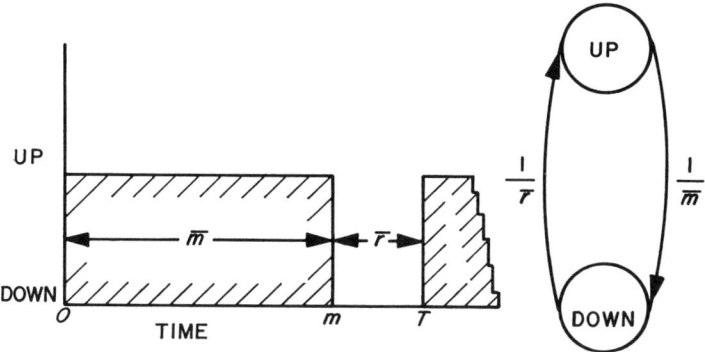

Figure 2.2 Two-state model.

service at the same time. Assume each is governed by a stationary renewal process with average times to failure of m_1 and m_2 and times to repair of r_1 and r_2 (see Figure 2.3).

Let the availability or long-term probability of success for the "series" circuit (in the electrical sense) be noted A_s. Since components 1 and 2 are independent, A_s may be determined from the product of the availabilities of the two devices, $A_1 \cdot A_2$:

$$A_s = A_1 \cdot A_2 = \frac{m_1}{m_1 + r_1} \cdot \frac{m_2}{m_2 + r_2}. \tag{2.7}$$

It is necessary to find an equivalent "series" component that performs similarly to the combination of devices 1 and 2; therefore, define m_s and r_s as follows:

$$A_s = \frac{m_s}{m_s + r_s}, \tag{2.8}$$

where m_s and r_s are the average uptime and downtime for the series equivalent.

The frequency of system failures is equal to the sum of the average frequency of the events of component 1 failing while 2 is operating plus the frequency of events of component 2 failing when 1 is up:

$$f_s = A_2 \cdot f_1 + A_1 \cdot f_2 = \frac{1}{m_s + r_s} \tag{2.9}$$

$$= \frac{m_2}{m_2 + r_2}\left(\frac{1}{m_1 + r_1}\right) + \frac{m_1}{m_1 + r_1}\left(\frac{1}{m_2 + r_2}\right). \tag{2.10}$$

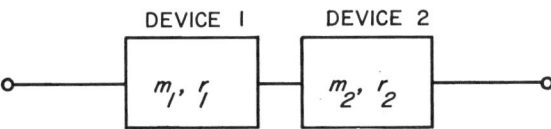

Figure 2.3 Series-connected components.

Note, the system cannot fail when it is already failed; hence failures of component 1 when 2 has already failed do not count as additional failures and conversely. The average up duration m_s or mean time-to-failure MTTF is readily determined from Equation 2.8 for A_s and Equation 2.10 for f_s, if one notes by the definition given in Equation 2.9 that

$$A_s = f_s \cdot m_s; \tag{2.11}$$

hence,

$$m_s = \frac{A_s}{f_s} = \frac{m_1 m_2}{m_1 + m_2}, \tag{2.12}$$

or

$$\frac{1}{m_s} = \frac{1}{m_1} + \frac{1}{m_2}. \tag{2.13}$$

The reciprocal of the mean time-to-failure is often designated as the failure rate λ. Thus, Equation 2.13 may be written as

$$\lambda_s = \lambda_1 + \lambda_2. \tag{2.14}$$

The failure rate of a "series" reliability system is the sum of the component or device failure rates.

The equivalent repair time r_s may now be solved by substitution of m_s from Equation 2.13 into Equation 2.7 and Equation 2.8:

$$r_s = \frac{1 - A_s}{f_s} \tag{2.15}$$

$$= \frac{(m_1 + r_1)(m_2 + r_2) - m_1 m_2}{m_1 + m_2}$$

The Renewal Process (Repairable System)

$$= \frac{\dfrac{r_1}{m_1} + \dfrac{r_2}{m_2} + \left(\dfrac{r_1}{m_1}\right)\left(\dfrac{r_2}{m_2}\right)}{\dfrac{1}{m_1} + \dfrac{1}{m_2}}$$

$$= \frac{\lambda_1 r_1 + \lambda_2 r_2 + (\lambda_1 r_1)(\lambda_2 r_2)}{\lambda_1 + \lambda_2} \tag{2.16}$$

$$= \frac{\lambda_1 r_1 + \lambda_2 r_2 + (\lambda_1 r_1)(\lambda_2 r_2)}{\lambda_s}. \tag{2.17}$$

As a further remark, if components 1 and 2 are dependent to the degree that 1 cannot fail while 2 is on repair, and vice versa, then Equation 2.17 becomes

$$r_s = \frac{\lambda_1 r_1 + \lambda_2 r_2}{\lambda_s}. \tag{2.18}$$

As a practical matter, the distinction is generally unimportant. The product λr is generally less than 0.1 for generation equipment and generally less than 0.01 for all transmission and distribution equipment; hence the difference between Equation 2.17 and Equation 2.18 is a second-order small quantity.

The form of Equation 2.18 has the satisfying appearance of a weighted average for the mean downtime.

EXAMPLE 2.1
A feeder is composed of an overhead section and an underground section. Find the failure rate and restoration time for the feeder given the following component rates:
Overhead feeder: 0.1 fault/cct-mi-yr.
Underground feeder: 0.1 fault/cct-mi-yr.
Cable termination: 0.002 fault/termination-yr.
The calculation is given in Table 2.1.

Table 2.1 Calculation for Example 2.1

Component	λ (per yr)	r (h)	λr (h/yr)	λr (yr/yr)
Overhead section (2 mi)	0.2	4	0.8	0.0000913
Underground section (1 mi)	0.1	24	2.4	0.0002740
Cable terminations (2)	0.004	4	0.016	0.0000018
Contribution to feeder	0.304		3.216	0.0003671

$r_s = \dfrac{3.216}{0.304} = 10.57 \text{ h}$

$\lambda_s = 0.304/\text{yr}$

$\bar{A}_s = 0.0003671$

$A = 0.9996329$

Note that the λr products (on a dimensionless basis) are extremely small compared to unity; hence, the difference between Equation 2.17 and Equation 2.18 would be negligible. Note further that the probability distribution of downtimes was not required to be exponential.

One note of caution needs to be included. The value r_s or average time out for the series circuit should not be assumed to be the mean of an exponential probability distribution even if r_1 and r_2 are associated with exponential distributions. The distribution of downtimes is not necessarily exponential. Further application of Equations 2.14 and 2.18 to distribution-system analysis are illustrated in Examples 2.2 and 2.3.

2.2.3 Repairable Components in Parallel

Suppose that either of two statistically independent devices can serve the load. Assume that each is governed by stationary failure and repair processes such that the distributions of downtimes and uptimes form a renewal process. The system is down when the outages of the two devices overlap. See Figure 2.4.

The behavior of this arrangement is essentially the *dual* of the series circuit. Where the hazard terms or failure rates were summed in the series circuit, the "repair rates" or reciprocals of downtimes are now summed to determine the system downtime.

The system unavailability is given by the product of the un-

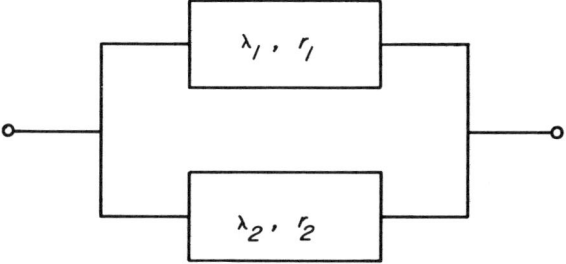

Figure 2.4 Parallel-connected components.

availability of component 1, $\lambda_1 r_1/(1 + \lambda_1 r_1)$, and component 2, $\lambda_2 r_2/(1 + \lambda_2 r_2)$:

$$\bar{A}_p = \frac{(\lambda_1 r_1)(\lambda_2 r_2)}{(1 + \lambda_2 r_2)(1 + \lambda_1 r_1)}. \tag{2.19}$$

The frequency of system failure is given by

$$f_p = f_1 \bar{A}_2 + f_2 \bar{A}_1$$

and

$$f_p = \left(\frac{\lambda_1}{1 + \lambda_1 r_1}\right)\left(\frac{\lambda_2 r_2}{1 + \lambda_2 r_2}\right) + \left(\frac{\lambda_2}{1 + \lambda_2 r_2}\right)\left(\frac{\lambda_1 r_1}{1 + \lambda_1 r_1}\right).$$

Simplifying leads to

$$f_p = \frac{\lambda_1 \lambda_2 (r_1 + r_2)}{(1 + \lambda_1 r_1)(1 + \lambda_2 r_2)}. \tag{2.20}$$

The equivalent element will have an average uptime m_p and the average downtime r_p. From Equations 2.5 and 2.6, $\bar{r}_p$ may be expressed in terms of $\bar{A}_p$ and f_p as follows:

$$r_p = \frac{\bar{A}_p}{f_p} = \frac{r_1 r_2}{r_1 + r_2}. \tag{2.21}$$

Table 2.2 Independent Events for More than Two Components

One of Three Components Out
Probability = $\bar{A}_1 \cdot A_2 \cdot A_3$

$$\text{Duration} = \frac{1}{\left(\dfrac{1}{r_1} + \dfrac{1}{m_2} + \dfrac{1}{m_3}\right)}$$

$$\text{Frequency} = \frac{\text{Probability}}{\text{Duration}}$$

Two of Three Components on Overlapping Outage
Probability = $\bar{A}_1 \cdot \bar{A}_2 \cdot A_3$

$$\text{Duration} = \frac{1}{\left(\dfrac{1}{r_1} + \dfrac{1}{r_2} + \dfrac{1}{m_3}\right)}$$

$$\text{Frequency} = \frac{\text{Probability}}{\text{Duration}}$$

From Equations 2.4, 2.19, and 2.21, m_p can be determined by substitution:

$$m_p = \frac{1 + \lambda_1 r_1 + \lambda_2 r_2}{(\lambda_1 \lambda_2)(r_1 + r_2)}. \tag{2.22}$$

From Equation 2.22 an equivalent failure rate λ_p may be defined as

$$\lambda_p = \frac{1}{m_p} = \frac{\lambda_1 \lambda_2 (r_1 + r_2)}{1 + \lambda_1 r_1 + \lambda_2 r_2}. \tag{2.23}$$

In a dual of the series-system case, the system uptime cannot be assumed to be exponentially distributed even if m_1 and m_2 are associated with exponential distributions. Generalization of these concepts to overlapping outages of one or more of several components is illustrated in Table 2.2.

The Renewal Process (Repairable System)

EXAMPLE 2.2
Two 75-MW hydrogenerators have identical forced outage characteristics: $\lambda_f = 0.00488$/day, $r_f = 1.066$ days.* What is the duration and frequency of occurrence of overlapping forced outages?

$$r_p = 0.533 \text{ day} = \frac{1}{1/1.066 + 1/1.066},$$

$$m_p = \frac{1 + (0.00488)(1.066) + (0.00488)(1.066)}{(0.00488)(0.00488)(1.066)(1.066)}$$

$$= \frac{1.0104}{10^{-4}(0.508)} = 19{,}900 \text{ days}.$$

$$\text{Frequency of occurrence} = \frac{1}{T} = \frac{1}{19{,}900 + 0.533} = 0.00005 \text{ per day}.$$

What is the mean duration and frequency of 150-MW capability (both units up)?
Note: Since both units must be up, this is equivalent to a "series" requirement.

$$\lambda_s = 0.00488 + 0.00488 = 0.00976/\text{day}.$$

$$\therefore m_s = 103.5 \text{ days},$$

$$r_s = \frac{2(0.00488)(1.066)}{0.00976} = 1.066 \text{ days},$$

$$f_s = \frac{1}{103.5 + 1.066} = 0.00965/\text{day},$$

$$T_s = 103.568 \text{ days}.$$

* These data were taken from the 1970 generating equipment report of the Statistical Analysis of System Outages Committee of the Canadian Electrical Association.

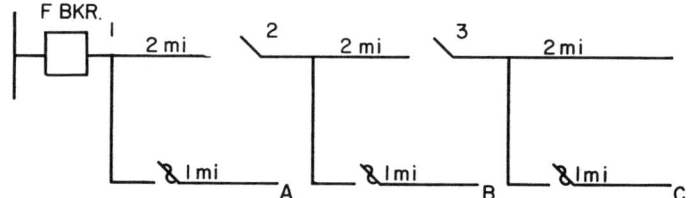

Figure 2.5 Diagram for Example 2.3.

EXAMPLE 2.3 MANUALLY SECTIONALIZED PRIMARY MAIN
See Figure 2.5 for the diagram and Table 2.3 for the interruption analysis of this example.

Primary main permanent fault characteristics:

0.07 fault/cct-mi-yr, 3-h av repair time.

Primary lateral faults resulting in fuse clearing:
0.18 fault/cct-mi-yr, 1-h av repair time.

Allow 0.5 h for manual sectionalizing.

EXAMPLE 2.4 LOOPED PRIMARY WITH MANUAL SECTIONALIZING
See Figure 2.6 for the diagram and Table 2.4 for the interruption analysis of this example. The following characteristics are assumed:
λ primary = 0.07 fault/cct-mi-yr,
r primary = 3 h av repair time,
λ lateral = 0.18 fault/cct-mi-yr,
r lateral = 1 h av repair time.
Allow 0.5 h for manual sectionalizing.

2.3 Models for Redundant Component Overlapping Outages

A probability model was developed in the previous section for the independent overlapping outage of redundant components. The expression for the rate of overlapping outage was given as Equation 2.23 and the duration of overlapping outage given as Equation 2.21. Additional models have been derived to recognize specific types of operating experience that cannot be properly recognized by the model derived previously. Four of these additional models considered in the following

Models for Redundant Component Overlapping Outages

Table 2.3 Interruption Analysis for Example 2.3

Component	λ (per yr)	Customer on Lateral A		Customer on Lateral B		Customer on Lateral C	
		r (h)	λr (h/yr)	r (h)	λr (h/yr)	r (h)	λr (h/yr)
Primary main 0.07 fault/cct-mi-yr							
Sec. 1 2 mi × 0.07 = 0.14		3	0.42	3	0.42	3	0.42
Sec. 2 2 mi × 0.07 = 0.14		0.5	0.07	3	0.42	3	0.42
Sec. 3 2 mi × 0.07 = 0.14		0.5	0.07	0.5	0.07	3	0.42
Primary lateral 0.18 fault/cct-mi-yr							
1 mi × 0.18 = 0.18		1	0.18	1	0.18	1	0.18
Sustained interruption rate = 0.60			0.74		1.09		1.44

(Due to primary main and lateral only) $r_A = 1.2$ h, $r_B = 1.8$ h, $r_C = 2.4$ h

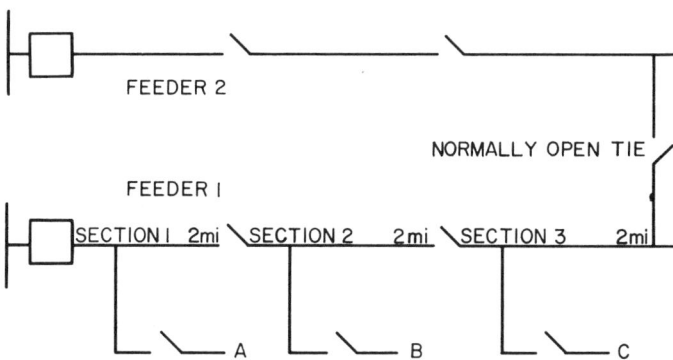

Figure 2.6 Diagram for Example 2.4.

Table 2.4 Interruption Analysis for Example 2.4. First contingency only (neglect risk of overlapping feeder outages)

Component	λ (per yr)	Customer on Lateral A		Customer on Lateral B		Customer on Lateral C	
		r (h)	λr (h/yr)	r (h)	λr (h/yr)	r (h)	λr (h/yr)
Primary main							
Sec. 1 2 mi × 0.07 = 0.14		3	0.42	0.5	0.07	0.5	0.07
Sec. 2 2 mi × 0.07 = 0.14		0.5	0.07	3	0.42	0.5	0.07
Sec. 3 2 mi × 0.07 = 0.14		0.5	0.07	0.5	0.07	3	0.42
Primary lateral							
1 mi × 0.18 = 0.18		1	0.18	1	0.18	1	0.18
Sustained interruption rate = 0.60			0.74		0.74		0.74

(Due to primary main and lateral only) $r_A = 1.2$ h, $r_B = 1.2$ h, $r_C = 1.2$ h

section are forced outage events occurring during maintenance periods, failure bunching due to environmental effects, overload induced outages of a second component during the forced outage of the first component, and common-mode failure events.

A good example of nonindependence for transmission circuit outages arises in the lightning-caused outage of both circuits supported on the same towers. The risk of outage of both circuits is appreciably larger than would be predicted by use of Equation 2.19 under the assumption of independence.

2.3.1 Maintenance Effects

First consider the forced outage events that occur during maintenance periods. This phenomenon is illustrated in Figure 2.7.

This model assumes that during a maintenance outage of one of two redundant components it is possible that the second component may suffer an unscheduled or forced outage. The model is restricted in that a maintenance outage would not be permitted during the duration of forced outage of the other component and therefore system interruption

Models for Redundant Component Overlapping Outages 23

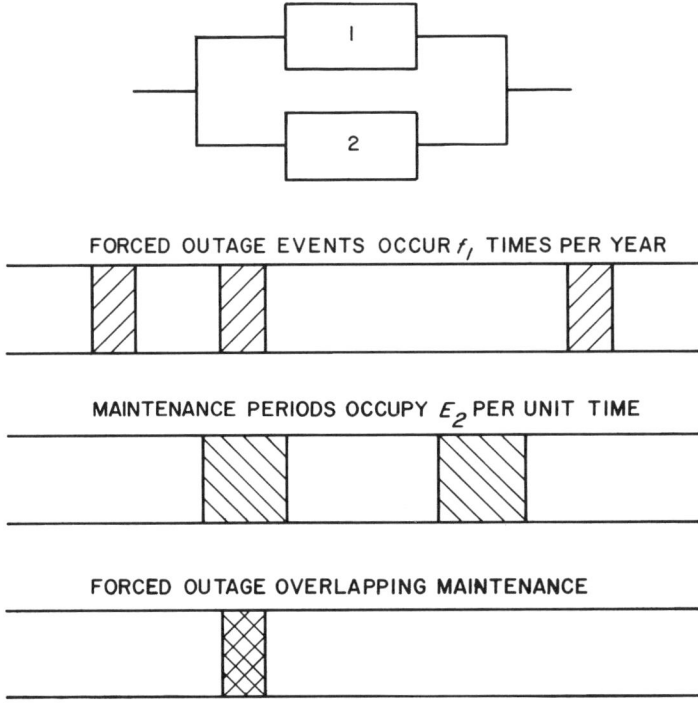

Figure 2.7 Forced outage events occurring during maintenance periods.

would not be initiated by maintenance. A maintenance outage on one of the components, say component number 2, is assumed to occur E_2 per unit time. If forced outage events on the first component occur f_1 times per year, then forced outage events on the first component can occur during maintenance outage of the second component at a rate very nearly equal to $E_2 \cdot f_1$ overlapping events/year.* The value of E_2 is estimated from an accumulation of the required scheduled outage time for maintenance and overhaul per unit of period time such as a

---

* The overlapping outage rate is given by the product of the probability that component number 2 is on maintenance outage, given that component number 1 is available, times the failure rate for component number 1. The approximation, "$E_2 \cdot f_1$," assumes that the product of failure rate times average repair time for component number 1 is very small compared to unity, an excellent assumption for transmission components.

year. For example, consider parallel power transformer banks in which the failure rate for a bank is 0.012 failure per year and in which each transformer bank is maintained approximately 35 hours per year. Assume the banks are redundant such that service continuity is achieved provided at least one bank is operating. Then the risk of transformer bank 1 failing during maintenance outage of transformer bank 2 will be equal to the product of the failure rate 0.012 times the exposure period 35 hours in the 8760 hours; $E_2 f_1 = (0.012)(35/8760) = 0.000048$/year overlapping outages of bank 1 during maintenance outage of bank 2. The forced outage of either bank during maintenance of the other bank will be twice this figure since there are two ways that the station interruption event could occur.

In summary, if λ_1, r_1, and f_1 are the average failure rate, repair time, and outage frequency for component number 1 and if E_2 and r_m are the per-unit time and average duration of maintenance outage for component number 2, then forced outage overlapping maintenance occurs at a rate of

$$\left[\frac{1 + \lambda_1 r_1}{1 + (\lambda_1 r_1)/(1 + r_1/r_m)} \right] f_1 \cdot E_2 \qquad (2.24)$$

times per year.

2.3.2 Failure Bunching

Failure bunching, that is, the greater risk of overlapping forced outage during periods of high environmental stress, has been treated in the published literature.[5,6] This phenomenon is illustrated in Figure 2.8, where the periods of severe weather labeled S are followed by periods of normal weather labeled N. Severe weather periods are thus assumed to recur on the average once in $N + S$ hours and to persist for S hours. During the period of normal environment the component failure rate is assumed to be λ failures per unit time. During the severe weather period of time, the component failure rate rises to λ'. The ratio of severe weather to normal weather failure rates is designated by ρ.

The average failure rate for the device λ_e is equal to the weighted average of the storm and normal failure rates and when dealing with a series system this average failure rate λ_e is used. When dealing with redundant systems, however, recognition must be given to the effect of

Models for Redundant Component Overlapping Outages

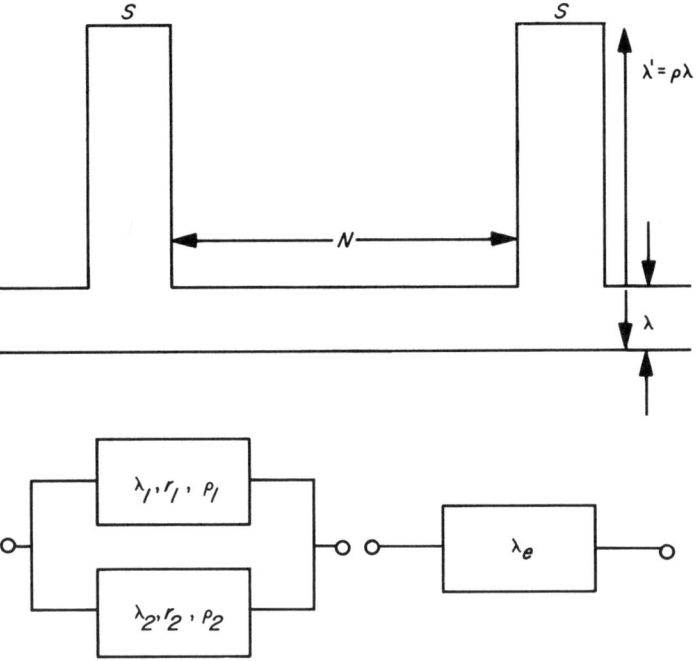

Figure 2.8 Failure bunching.

short periods of high environmental stress that can cause an increase in the risk of overlapping forced outage. Failure bunching effects are illustrated for two devices with normal weather failure rates λ_1 and λ_2, average repair times r_1 and r_2, and ratios of storm or severe weather failure rate to normal weather failure rate ρ_1 and ρ_2, respectively. The rate of overlapping failure for two devices exposed to a severe weather and a normal weather environment is given as Equation 2.26.

$$\lambda_e = \frac{(S\lambda' + N\lambda)}{N + S} = \frac{(\rho S + N)\lambda}{N + S}, \qquad (2.25)$$

$$\lambda_p = \frac{N}{N + S} \lambda_1 \lambda_2 \left\{ (r_1 + r_2)\left[1 + \frac{S}{N}(\rho_1 + \rho_2)\right] + 2\frac{S^2}{N}\rho_1\rho_2 \right\}, \qquad (2.26)$$

$$\lambda_p > \lambda_{e_1}\lambda_{e_2}(r_1 + r_2). \tag{2.27}$$

Note that additional factors appear in Equation 2.26 as compared to the independent constant-failure-rate model that was developed earlier and illustrated in Equation 2.27. To illustrate the influence of failure bunching, consider two 20-mile-long circuits feeding a substation in which the average permanent failure rate is one failure per year on each circuit and the average downtime is 12 hours. Assume that the circuits are over separate rights-of-way so that they may be considered to be independent and that on the average the storms persist for 2 hours and recur on the average once every 200 hours during the summer season. A summer season redundant system overlapping outage rate can be estimated using Equations 2.25 through 2.27. Assume that during the severe weather period the failure rate rises to 50 times the normal weather failure rate. The calculations comparing the effects of environmental stress are illustrated in Table 2.5. For the example developed, the increase in over-

Table 2.5 Calculations for Environmental Stress

$$\lambda_e = 1 \text{ permanent failure/yr}$$
$$r = 12 \text{ h}$$
$$N + S = 200 \text{ h}$$
$$S = 2 \text{ h}$$
$$\rho = 50$$
$$\therefore \lambda = \lambda_e \left(\frac{N+S}{N+\rho S}\right) = (1)\left(\frac{200}{298}\right) = 0.668/\text{normal yr}$$
$$\lambda' = \rho\lambda = 33.4/\text{yr of stormy weather}$$

From Equation 2.26

$$\lambda_p = \left(\frac{198}{200}\right)(0.668)^2 \left\{\left(\frac{12+12}{8760}\right)\left[1 + \frac{2}{198}(50+50)\right]\right.$$
$$\left. + \frac{2 \times 2^2}{8760 \times 198} \times 50^2\right\}$$
$$= 0.0075/\text{yr}$$

From Equation 2.27

$$\lambda_{e_1}\lambda_{e_2}(r_1 + r_2) = (1)(1)\left(\frac{12+12}{8760}\right) = 0.0029/\text{yr}$$

Models for Redundant Component Overlapping Outages

lapping failure rate due to failure bunching caused by environmental stress represents a better than 2.6 : 1 increase.

2.3.3 Overload Effects

Another factor to consider in conjunction with redundancy is the risk of component overload during forced outage or maintenance outage of other components. Consider the parallel transformer arrangement shown in Figure 2.9. When one bank is out of service, there is a risk

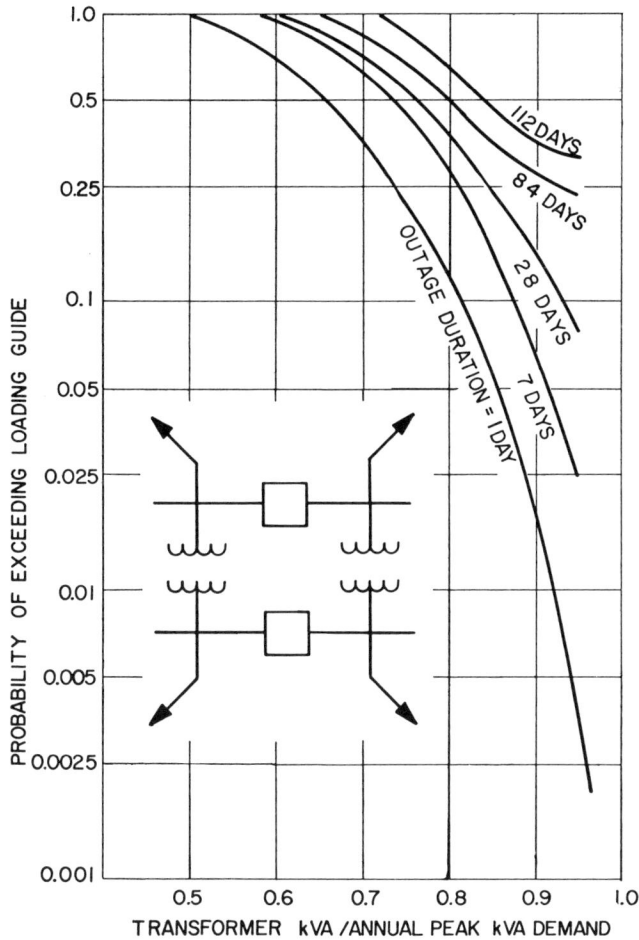

Figure 2.9 Overload contingency curves for an FOA transformer.

that the other bank may not be able to carry the daily load and remain within the recommended limits set by the loading guides. Suppose, for example, that one transformer has been forced out of service for a period of time. Depending upon the seasonal load and ambient temperature and upon the capacity of the remaining bank in service, there is a risk that the daily load cycle imposed upon the remaining transformer will exceed its capability measured in terms of the recommendations in ANSI Appendix C57.92, June 1962, "Guide for Loading Oil Immersed Distribution and Power Transformers." Figure 2.9 shows the risk that an outage on one transformer that was initiated at random throughout the year will result in loads upon the remaining transformer that exceed the guide recommendations. These curves were developed for typical northeast U.S.A. load and temperature cycles. For example, if the outage duration is 1 day and the capacity of the transformer is 0.8 of the annual peak kVA demand, then there is about a 12% risk that the day selected at random will result in load and ambient temperature conditions that will exceed the guide recommendation. In this specific example, the combined bank kVA is 2 × 0.8 (or 1.6) times the annual kVA demand. As the outage duration lengthens, the risk of incurring a day in which the load and ambient conditions will exceed the guide recommendation increases. If a long duration outage occurs, such as one of 84 days, then the risk of encountering at least 1 day in which the load and ambient temperature conditions would exceed the loading guide for a transformer whose rating was 0.8 times the annual peak kVA demand rises to more than 50%. Now consider the risk that a

Table 2.6 Overload Event Rate

$f_0 = \lambda_T P[R] P[O \mid R]$

Example—1-wk outage

Transformer kVA = $\frac{2}{3}$ annual peak

$\lambda_T = 0.012/\text{yr}$

$P[R] = 0.54$

$P[O \mid R] = 0.75$

$f_0 = (0.54)(0.75)(0.012) = 0.00487/\text{yr}$

$f_s = 2f_0 = 0.00974/\text{yr}$

Models for Redundant Component Overlapping Outages   29

forced outage of one transformer bank will occur during a period of time that the load and ambient temperature conditions will exceed the recommendations of the loading guide on the remaining transformer bank. An example is worked out in Table 2.6 for the overload event rate. Three factors are involved: first, the unscheduled outage rate for transformers λ_T; second, the distribution of outage durations for transformers $P_r[R]$; third, the probability that an overload will result, given the outage duration, $P[O \mid R]$. Assume a transformer failure rate of 0.012 failure per year and consider an arrangement of two parallel transformer banks with each transformer having a capacity equal to two-thirds of the annual peak. Consider the case of a 1-week outage duration. Based upon outage duration data, the chance of a 1-week outage was estimated to be about 54%.

From Figure 2.9, the risk of overload is approximately 75% for an outage of 7 days for a transformer bank with nameplate kVA equal to two-thirds of the annual peak kVA demand. The frequency with which outage events on one bank will lead to overload conditions on the second bank are calculated and illustrated in Table 2.6. Here the calculations are illustrated only for the event of a one-week outage, given the probability of that outage duration is 0.54. A complete tabulation of the calculation is illustrated in Table 2.7.

Overload risk curves as illustrated in Figure 2.9 may be determined by simulation using forecasts of day-by-day loads and ambient temperature conditions. For specified time of outage initiation and outage duration, a check can be made against the loading guide to see whether or not the outage would result in loads on the remaining transformer that will exceed the guide. From a sequence of such studies, an estimate of the risk that an outage initiated at random and lasting for a specified duration would result in loading conditions exceeding the loading guide for the remaining transformer bank can be prepared. The curves illustrated in Figure 2.9 were developed for northeast U.S.A. seasonal load types and northeast U.S.A. seasonal ambient temperatures. These curves were for transformers with thermal characteristics similar to FOA type transformers.

2.3.4 Common-Mode Failures

The fourth aspect to consider in redundant systems is the possibility of systematic nonrandom events causing an outage of all the redundant

Table 2.7 Transformer Bank Outage Events Resulting in Overload on the Remaining Transformer

Twin transformer bank rating = $\frac{4}{3}$ annual peak demand

Duration of outage	"R" (days)	1	7	28	84	112	Total
Outage duration (probability density function)	$P[R]$	0.402	0.540	0.0416	0.0082	0.0082	1.00
Outage event (occurrence frequency per thousand years)	$f_e(R) = 12\, P[R]$	4.8	6.5	0.5	0.1	0.1	12.0
Probability of at least one "overload day" during outage event	$P[O \mid R]$	0.47	0.75	0.82	0.97	1.0	—
Frequency of outage (events leading to overload per thousand years)	$f_o = f_e(R)P[O \mid R]$	2.26	4.87	0.41	0.097	0.1	7.74
Exposure (2 units)	E	2	2	2	2	2	2
Frequency of station events leading to overload per thousand years	$f_s = E f_o$	4.52	9.74	0.82	0.194	0.2	15.48

components. These events have been called common-mode failures. Examples of common-mode failures include cases of all separate but identical components containing the same built-in design error, unrecognized dependence of all redundant components on a single common element, systematic human error, changes in the characteristics of the system, and changes in the environment.

Efforts have been directed to develop procedures to mitigate the effect of systematic nonrandom failures that occur in control and protection systems for nuclear reactors.[3] The system design must include a sufficient degree of functional redundancy so that the common-mode failure of a particular type of protective or control function will not prevent the safe shutdown of the reactor. That is, alternative means must be available to take over on the loss of the particular protection or control function so that shutdown can be achieved without a serious disturbance or serious hazard. Very little data are available on the rate with which common-mode events have occurred. The fact that they can and have occurred is sufficient to recommend consideration of tests and inspection procedures to discover the onset of common-mode failure events in critical control and protection systems.

A summary of common-mode failure experience with reactor protection and control systems has been offered by Eppler,[7] who came to the following interesting conclusions: (a) Common-mode hazards proceeded to failure in three cases out of 300 subsystem years, indicating a common-mode rate of 0.01 per subsystem year. In seven cases, the trouble was arrested before failure had occurred, and in four additional cases the potential for trouble was discovered and corrected before damage occurred. (b) If two redundant units provide a system with lower probability of failure from independent randomly occurring overlapping outages than the common mode, then there is no need for three such units except for coincidence and/or convenience.

I. M. Jacobs classifies common-mode failures, CMFs, as Functional Deficiency (misapplication of monitoring equipment or mismodeling of system response), Maintenance Error, Design Deficiency, or External Event.[3] Jacobs notes five categories of corrections for CMFs: Functional Diversity, Operational Administrative Diversity, Design Administrative Diversity, Equipment Diversity, and Physical Diversity.

2.4 Reliability Procedures for Substations

A logical place to start in developing a review of substation reliability evaluation procedures is to consider an analysis of the types of failures that have occurred in substations. For example, from FPC Reports[8] of 200 "disturbances" 39 were associated with equipment difficulties in substations and 12 were associated with errors in operation. Eleven of the equipment difficulties were associated with protection, control, and supervisory equipment. Statistics such as these serve to point out the relative frequency of occurrence of disturbances initiating in certain portions of an electric power system and, in particular, in substations. It is of interest to speculate on the relationship between the "disturbances initiated" and the dollar investment in various systems in the substation. For example, more than one-quarter of the substation

Reliability and Availability Applications to Distribution Systems   32

equipment difficulties were associated with protection, control, and supervisory functions. Such statistics, therefore, may be of value to designers to aid in review of design practices and to operators to aid in review of maintenance and testing practices as well as in training given personnel for testing and maintaining substations.

2.4.1 Basic Procedure for Substation Evaluation

A basic procedure for substation evaluation is outlined and subsequently illustrated by a small example. There are five essential steps to the procedure.

2.4.1.1 Physical System Description Specify the component and circuit ratings, impedances, and connections within the boundary of study. Specify circuit and component outage modes, rates, repair statistics, and maintenance requirements, that is, per-unit time out for maintenance and average duration of maintenance outage. Specify supply and load terminals.

2.4.1.2 Performance Criteria Specify the performance criteria for successful system operation. The criteria may include component overload as well as system frequency and bus voltage limits, on the one hand, and circuit continuity, on the other.

2.4.1.3 Reliability Goal Establish a level of satisfactory system performance. Negatively the level can be expressed in terms of events that lead to the system not meeting the performance criteria. Positively, the level may be set qualitatively in terms of the contingency level (number of overlapping outage events) the system can withstand and yet meet the performance criteria. The level may also be set numerically in terms of the reliability of system performance measured in terms of the interval of time between events leading to system failure or in terms of the per-unit time the system meets the performance criteria, that is, the "availability" of the system.

2.4.1.4 Failure Modes and Effects Analysis Decide on the sequence of failure events and the level of contingencies to be investigated. See Table 2.8.

Table 2.8 Failure Modes and Effects Analysis

Component identification	Failure effect
Component function(s)	Failure consequence: system effect
Component failure modes	Detection means
Failure environment and time	Circumvention means

At this step, specify the conditions of system load and the state of component maintenance to be investigated with the failure events. Given the load and maintenance state, the effect of the failure event may be checked as follows:

1. Determine the effect of the failure event upon the protective system, and determine the resulting breaker action.
2. Determine the effect of the breaker action upon the power system. Determine if overload or out-of-limit bus voltage or load interruption has occurred.
3. Determine if the performance criteria have been violated.
4. If the performance criteria have not been met, determine what steps are required to bring the system back to success state:
  a. Automatic transfer possible?
  b. Manual transfer possible?
  c. Component and circuit tests and repairs required?
5. Record the effect of the failure event by the terminals affected; that is, for each affected terminal, store the failure mode, the event probability, and the event duration. Refer to Table 2.9 for a summary of formulas for computing event probability, duration, and frequency.

2.4.1.5 Accumulation of Failure Effects and Summary Prepare a list of failure events that lead to violation of the performance criteria. Order the list by event probability (or event frequency) to expose the events that by prediction are events expected to cause the most trouble. Combine the system failure-event probabilities and frequencies by the rules for combining like capacity states. Since each failure event investigated can be described by an exclusive state, the probability of occurrence of the system failure state is equal to the sum of the probabilities of the failure events. The probability of the failure state is equal to the product of the frequency of occurrence of the failure event and the associated duration of the failure state. For nonconsecutive failure states, the

Table 2.9 Summary of Formulas for Event Probability, Frequency, and Duration for Four Independent Renewal Processes

Parameters for each renewal process:
Up state availability : $A_i = m_i/(m_i + r_i)$
Up state duration : m_i
Outage state duration : r_i
$m_i \gg r_i$ for Transmission and Distribution Components

Basic independent events—four-component examples:

All Components Up	Component 1 Out	Components 1, 2 Out
$P_0 = A_1 A_2 A_3 A_4$	$P_1 = (1 - A_1) A_2 A_3 A_4$	$P_{12} = (1 - A_1) \times (1 - A_2) A_3 A_4$
$1/M_0 = 1/m_1 + 1/m_2 + 1/m_3 + 1/m_4$	$1/M_1 = 1/r_1 + 1/m_2 + 1/m_3 + 1/m_4$	$1/M_{12} = 1/r_1 + 1/r_2 + 1/m_3 + 1/m_4$
$f_0 = P_0/M_0$	$f_1 = P_1/M_1$	$f_{12} = P_{12}/M_{12}$

Events occurring during an exposure period—single-contingency example:
Let E be the per-unit exposure time and r_e be the average duration of the exposure period. Assume all components are operating and *only* component 1 fails during the exposure period leading to an interruption:

$f_{1e} = E f_1$.

The duration of the interruption event depends upon the means for restoration:

a. Restoration by automatic transfer:
   Set interruption duration equal to the switching time s.
   (Note: Component 1 is still in an outage mode.)
b. Restoration by repair of component 1:
   Set interruption duration equal to M_1.
c. Restoration by repair of component 1 or by completion of exposure period:
   Set interruption duration equal to $M_1 r_e/(M_1 + r_e)$.

Note: Only the effect of the single contingency has been accounted for; the overlapping double-contingent events must be investigated separately.

frequency of the sum of the failure states is equal to the sum of the frequencies of occurrence of the separate states.[9,10] A small, usually negligible, pessimistic error results if all the failure states are assumed to be nonconsecutive.

The system availability and/or reliability prediction may now be compared with the goal. Should the prediction not measure up to the goal, one logical procedure to find improvement is to look for ways to mitigate the causes underlying the events that contributed most to failure.

2.4.2 Substation Reliability Prediction

2.4.2.1 Physical System Description Consider the ring-bus configuration shown in Figure 2.10. Assume power is supplied over two widely separated lines, L_1 and L_2, such that the supply and line capabilities are fully redundant. The lines are assumed to perform as statistically independent components, and environmental effects are neglected. Lines S_1 and S_2 feed a multiple radial subtransmission network.

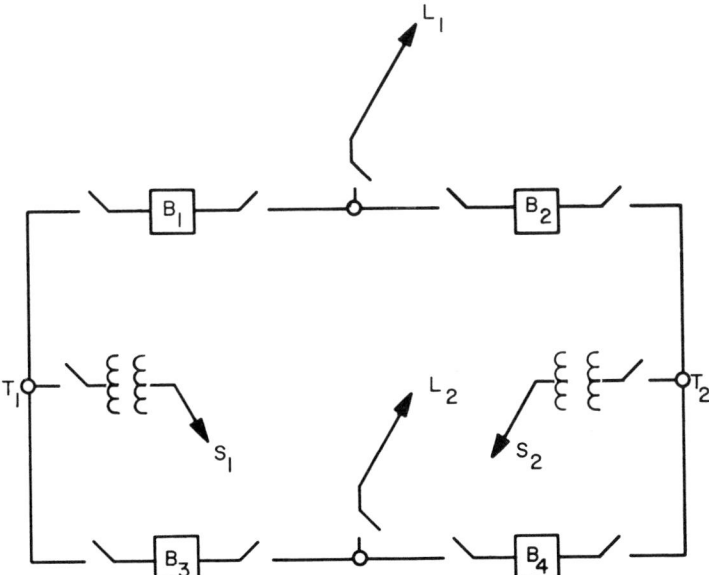

Figure 2.10 Ring-bus example.

Reliability and Availability Applications to Distribution Systems 36

Subtransmission breaker and switch equipment are outside the example. The rating of each transformer is two-thirds the annual peak load of the station. The breaker ratings are assumed to be not limiting. Outage, repair, and maintenance data assumed for the example are shown in Table 2.10 and are patterned after the data reported in the paper by Mallard and Thomas.[11] Disconnect switch failure events have been assigned to the bus, breaker, or transformer switched by the device.

2.4.2.2 Performance Criteria Successful system operation is achieved (a) as long as either L_1 or L_2 can transmit power to either S_1 or S_2 and (b) as long as the load on the substation transformers does not exceed the levels specified in the transformer loading guides C57.92.[12]

2.4.2.3 Reliability Goal No goal was set for this example. The example will illustrate prediction of station interruptions.

Table 2.10 Component Outage Data

	Failure Rate (per yr)	Outage Duration (h)	Unavailability	Probability
Circuit Breakers				
Circuit breaker fault (backup required)	0.007	72	—	—
Maintenance	—	16	0.004	—
Probability of breaker found inoperative	—	—	—	0.0005
Transformer				
Forced outage	0.012	168	—	—
Maintenance	—	12	0.004	—
Bus Section				
Forced outage	0.007	3.5	—	—
Maintenance (combined with line)	—	18	0.003	—
Line Section				
Forced outage	0.05	23	—	—
Maintenance	—	15	0.005	—

2.4.2.4 Failure Modes and Effects Analysis

A failure modes and effects analysis was performed for single- and double-contingent (overlapping) events. All events that resulted in loss of the substation are listed later in Table 2.11. The first event listed in the table assumes that a breaker fault occurs, given that a stuck breaker condition already exists in the substation. For example, suppose breaker B_1 were to fail, given that breaker B_2 is inoperative at the time of failure. These conditions are shown in Figure 2.11 and in the following list.

Failure mode:
Shown in Figure 2.11.

Protection response:
Primary: Trip breakers 2,3: Circuits L_1, T_1.
Auxiliary: Trip breakers 1,4: Circuits L_1, T_2.

Effect:
Substation outage.

Duration:
Switching operation.

A signal would be developed to trip breakers B_3 and B_2; B_2 would not operate; hence B_4 would have to be tripped by backup. Breakers at the

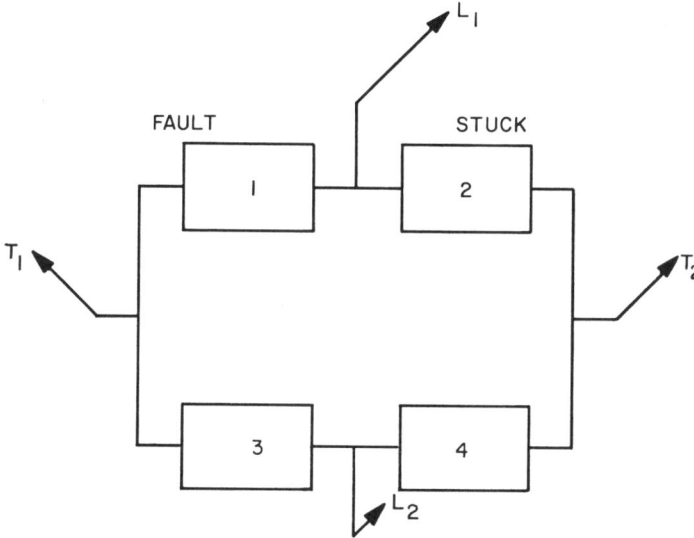

Figure 2.11 Ring bus with stuck breaker.

Table 2.11 Line, Transformer, Bus, Circuit Breaker Persistent Forced Outages and Maintenance Events That Lead to Station Interruption

Failure Event	Event Data					Station Effect		
	Combinations (1)	Exposure (2)	Restoration Means (3)	Duration (h) (4)	Freq. (per 1000 yr) (5)	Freq. (per 1000 yr) (6)	H (h/1000 yr) (7)	Freq. Rank (8)
I. Brkr. Fault/Stuck Brkr.	8	~1	S*	1	0.0035	0.028	0.028	8
II. Forced Outage/Maintenance on Another Component								
A. Line & Line Bus Maint.								
Line F.O /maint.	2	0.005	R†	9.1	50	0.5	4.52	1
Line bus F.O./maint.	2	0.005	R	2.8	7	0.07	0.196	6
Brkr. fault/maint.	4	0.005	S	1	7	0.14	0.14	2
B. Transf. & Bus Maint.								
Transf. F.O./maint.	2	0.004	R	11.2	12	0.096	1.07	5
Transf. bus F.O./maint.	2	0.004	R	2.7	7	0.056	0.152	7
Brkr. fault/maint.	4	0.004	S	1	7	0.112	0.112	3
C. Brkr. Fault/Brkr. Maint.	4	0.004	S	1	7	0.112	0.112	3

III. Overlapping, Independent Forced Outages

(Transf. + Bus)	2	~1	R	53.5	0.0044	0.0089	0.475
(Line + Bus)	2	~1	R	10.3	0.0076	0.0152	0.156
Breaker Opposite breakers	4	~1	S	1	0.0004	0.0016	0.0016
Adjacent breakers	8	~1	S	1	0.0000056	0.00004	0.00004
Breaker · Transformer	8	~1	S	1	0.0016	0.0129	0.0129
Breaker · Line	8	~1	S	1	0.00091	0.0073	0.0073
Bus · Breaker	16	~1	S	1	0.00002	0.0003	0.0003
Station interruption:						1.16	7.0
Average duration = 6.0 h							

* Restoration by switching.
† Restoration by repair.

Reliability Procedures for Substations 39

sending end of L_1 would also be tripped. Service to S_1 and S_2 can be restored by disconnecting B_1 and B_2 and supplying the transformers from line L_2. Assume 1 hour to analyze, test, and isolate B_1 and B_2. Note that column 4 in Table 2.11 lists restoration times for each failure event. With each breaker fault, there are two stuck breaker events that can cause the substation to be out. Accordingly, there are eight events of this type that can shut down the station, as is shown in column 1 of Table 2.11. The procedure for estimating the frequency of occurrences of this event is now given.

Breaker failure rate times conditional probability of stuck breaker = 0.007 × 0.0005.
For 8 combinations: Frequency—0.028/1000 years.

Strictly speaking, this frequency is a measure of the events per exposure year. It is assumed that a breaker will not fail when it is out for maintenance and that the station cannot fail if it has already failed. The hours of maintenance per year and the hours of failure per year should be deducted from the period hours in the year. Provision for the exposure has been incorporated in an exposure factor E shown in column 2 of Table 2.11. In this instance, the fraction of time out for maintenance and repairs is negligibly small compared to unity and has been neglected. The exposure factor E has therefore been set to unity. The product of the combinations (8), and exposure (1), and the event frequency (0.0000035) gives the station outage frequency of 0.000028 event per year or 0.028 event per thousand years, as shown in column 6. Column 7 of Table 2.11 lists the products of the event outage duration and the event frequency. Strictly speaking, this value should be divided by the factor $(1 + H/8760000)$ to give the expected hours out per thousand years.

Column 8 in Table 2.11 lists the ranking of contributions to unreliability (frequency). The breaker fault given a stuck breaker is eighth in line of contributors.

The net effect of all the modes of failure tabulated on Table 2.11 is shown to be a station outage every 860 station years with an average duration of outage of 6.0 hours. The most likely events to lead to station interruption are line, breaker, and transformer outage during a maintenance outage of a line or transformer. Continuing the failure modes and effects analysis, another potential failure event is a trans-

Reliability Procedures for Substations   41

former overload during forced outage of the other transformer. It was assumed that maintenance of transformers would be done during times of sufficiently reduced demand to permit the load to be carried on one unit without violation of the ANSI loading guide (assumed performance criteria). An example of an approach to estimate the risk of transformer overload during forced outage of the other unit is illustrated in Table 2.7. A distribution of outage times $P_r[R]$ (the second row in the table) has been assumed that gives the most likely outage duration to be 1 week and the average outage duration to be 1 week. The frequency of occurrence of outage events of given duration is shown in the third row. The probability of at least one overload for the given outage duration $P[O \mid R]$ is read from Figure 2.9 at a ratio of transformer kVA to annual peak kVA demand equal to 0.667. The exposure to this type of event is nearly unity on each transformer; hence an exposure factor of 2 has been assumed for the station.

2.4.2.5 Accumulation of Failure Effects and Summary For the ring-bus example, the two performance criteria resulted in widely disparate results. Station interruption, on the one hand, gave a predicted reliability of 1 interruption per 860 substation years. Component overload, on the other, gave a predicted reliability of 1 event per 65 substation years. Care is required before combining these measures into one composite value. The overload criteria may not necessarily imply an interruption if the operator elects to take transformer overload and consequent aging.

2.4.2.6 Industrial Substation Example Egly and Esser have shown the benefits of a reliability review for industrial substations.[13] The effects of split buses and redundant feeds are clearly shown in Examples 2.5 and 2.6. Failure data were taken from Patton,[14] from Mallard and Thomas,[11] and from Egly and Esser.[13] Useful failure-rate information on power-system components is available in a wide variety of publications.[15-25]

The two cases are offered to compare the station failure rate and expected outage hours for the single-source, single-low-voltage-bus station with a station with a double source and a single low-voltage bus with a tie breaker.

In Case A, the feeder-outage risk dominates both the station failure-rate causes and the expected outage hours per year. Only the single-contingency events leading to station interruption of the 13.8 kV supply are summarized. These events are of prime significance compared to the second-contingency and higher-order events.

In Case B, the single-contingency events leading to interruption of the 13.8 kV supply are fewer in number and contribute roughly one-tenth the failure rate and about one-sixth the expected outage hours per year. To achieve this improvement requires the additional 46 kV feed, three 46 kV breakers, and two 13.8 kV breakers.

EXAMPLE 2.5 INDUSTRIAL SUBSTATION CASE A
See Figure 2.12. Investigate cases of total loss of 13.8 kV bus. See Table 2.12.

EXAMPLE 2.6 INDUSTRIAL SUBSTATION CASE B
See Figure 2.13. (Assume that transformer and breaker-bus are fully redundant.) Investigate cases of total loss of 13.8 kV bus (1). See Table 2.13.

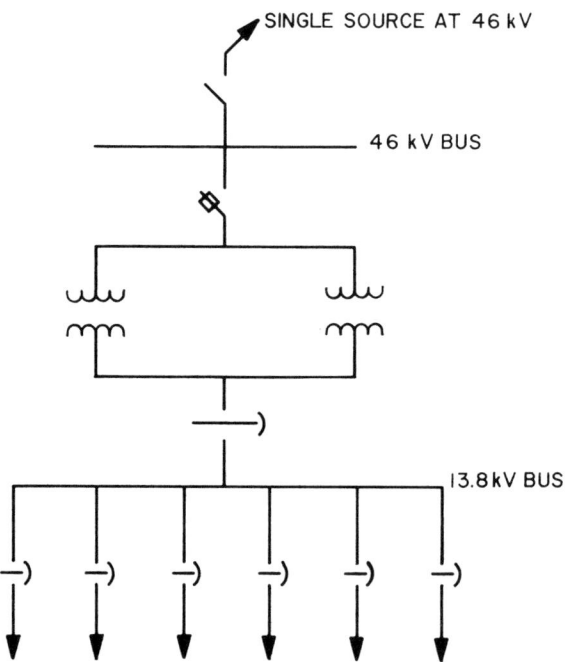

Figure 2.12 Industrial substation Case A.

Reliability Procedures for Substations

Table 2.12 Failure Rates for Industrial Substation Case A

Component	Failure Rate λ_c (per yr)	Repair Time r_c (h)	Number n_c	Station Failure Rate $n_c\lambda_c$ (per yr)	Station Expected Outage $n_c\lambda_c r_c$ (h/yr)
46 kV bus	0.002	1.2	1	0.002	0.0024
46 kV disconnect	0.001	1.5	2	0.002	0.003
46 kV transformer	0.004	5*	2	0.008	0.040
13.8 kV bus CB	0.010†	3.5	1	0.010	0.035
13.8 kV bus (enclosed)	0.002	1.2	1	0.002	0.0024
13.8 kV fdr. brkr. (fault)	0.002	3.5	6	0.012	0.042
Station caused 13.8 kV outages (subtotal)				0.036	0.1248
46 kV feeder	0.015/mi-yr	1.3	10 mi	0.15	0.195
Total				0.186	0.3198

* Time to disconnect faulted unit.
† Rate for faults and all unscheduled outages.

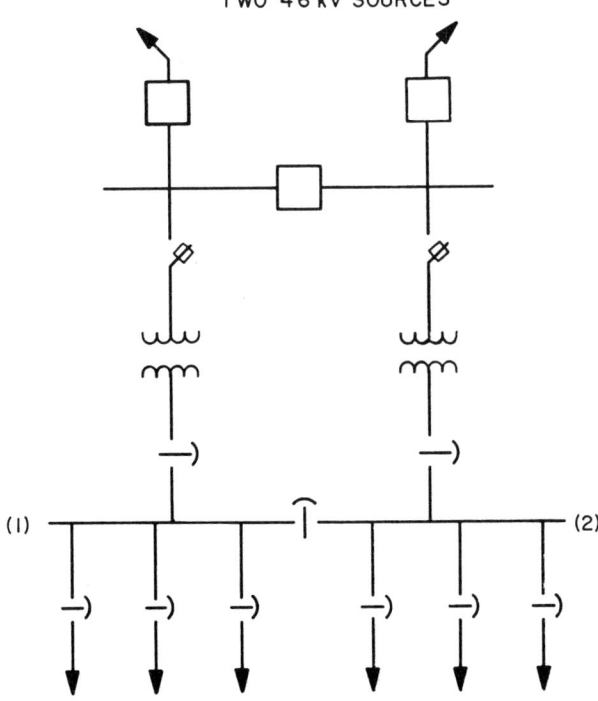

Figure 2.13 Industrial substation Case B.

Reliability and Availability Applications to Distribution Systems 44

Table 2.13 Failure Rates for Industrial Substation Case B

Component	Failure Rate λ_c (per yr)	Repair Time r_c (h)	Number n_c	Station Failure Rate $n_c\lambda_c$ (per yr)	Station Expected Outage $n_c\lambda_c r_c$ (h/yr)
Single Contingency Events					
46 kV OCB bus tie	0.005	2.4	1	0.005	0.012
13.8 kV bus CB and tie CB	0.002	3.5	2	0.004	0.014
13.8 kV fdr. brkr.	0.002	3.5	3	0.006	0.021
13.8 kV bus	0.002	1.2	1	0.002	0.0024
				0.017	0.0494

Single Contingency during Maintenance

46 kV cct out during maintenance of other cct. (Assume 2 12-h outages for maintenance each year.)

$$2 \times \frac{(2 \times 12)}{8760} \times 0.15 \; @ \; 1.3\text{h} \quad\quad 0.0008 \quad 0.0010$$

Total   0.018   0.050

Neglect double contingent events as second-order small; common-caused 46 kV circuit outages have been neglected.

References

1. P. H. Fowler, "System Pathology," *IEEE, Transactions on Reliability*, vol. 17, 1968, pp. 122–126.

2. A. E. Green, "Reliability Prediction," *Proceedings of the Institution of Mechanical Engineers (London)*, vol. 184, pt. 3B, 1970, pp. 17–24.

3. I. M. Jacobs, "The Common Mode Failure Study Discipline," *IEEE, Transactions on Nuclear Science*, vol. 17, no. 1, 1970, pp. 594–598.

4. J. Green, "Systematic Design Review and Fault Analysis," Institution of Mechanical Engineers Conference on Safety and Failure of Components, Paper 29, 1969, pp. 208–214.

5. D. P. Gaver, F. E. Montmeat, and A. D. Patton, "Power System Reliability I—Measures of Reliability and Methods of Calculation," *IEEE, Transactions on Power Apparatus and Systems*, vol. 83, 1964, pp. 727–737.

6. F. E. Montmeat, J. Zemkoski, A. D. Patton, and D. J. Cummings, "Power System Reliability II—Applications and a Computer Program," *IEEE, Transactions on Power Apparatus and Systems*, vol. 84, 1965, pp. 636–643.

References

7. E. P. Eppler, "Common Mode Failure Considerations in the Design of Systems for Protection and Control," *Nuclear Safety*, vol. 10, no. 1, 1969, pp. 30–45.

8. FPC Report—"Prevention of Power Failures," vol. 1, June 1967.

9. R. J. Ringlee and S. D. Goode, "On Procedures for Reliability Evaluations of Transmission Systems," *IEEE, Transactions on Power Apparatus and Systems*, vol. 89, 1970, pp. 527–536.

10. F. W. Davenport, E. M. Magidson, and Y. A. Yakub, "Substation Bus, Switching Arrangements—Their Essential Requirements and Reliability," *Electra* (CIGRE Publication), no. 10, October 1969, pp. 37–53.

11. S. A. Mallard and V. C. Thomas, "A Method for Calculating Transmission System Reliability," *IEEE, Transactions on Power Apparatus and Systems*, vol. 87, 1968, pp. 824–833.

12. "Guide for Loading Oil-Immersed Distribution & Power Transformers," ANSI Appendix C57.92, June 1962.

13. D. T. Egley and W. F. Esser, "Reliability Analysis and What It Means," *IEEE, Transactions on Industry and General Applications*, vol. 5, 1969, pp. 578–581.

14. A. D. Patton, "Determination and Analysis of Data for Reliability Studies," *IEEE, Transactions on Power Apparatus and Systems*, vol. 87, 1968, pp. 84–99.

15. Edison Electric Institute, "Cable Operation 1963," EEI Publication 65-36.

16. W. H. Dickinson, "Report on Reliability of Electric Equipment in Industrial Plants," *AIEE Transactions*, vol. 81, pt. II, 1962, pp. 132–151.

17. Edison Electric Institute, "EEI Report on Transformer Troubles 1968," EEI Publication 70-29.

18. G. L. Landgren, "Evaluation of Distribution Systems for Medium Load-Density Commercial Areas," *AIEE Transactions*, vol. 77, pt. III, 1958, pp. 128–137.

19. "Report of Joint AIEE-IEEE Subject Committee on Line Outages," *AIEE Transactions*, vol. 71, pt. III, 1952, pp. 43–55.

20. IEEE-EEI Committee Report, "EHV Line Outages," *IEEE, Transactions on Power Apparatus and Systems*, vol. 86, 1967, pp. 547–562.

21. AIEE Committee Report, "Report of Field Experience with Aerial Power Cable," *AIEE Transactions*, vol. 78, pt. III, 1959, pp. 1688–1698.

22. R. A. W. Connor and R. A. Parkins, "Operational Statistics in the Management of Large Distribution Systems," *Proceedings of the Institution of Electrical Engineers* (*London*), vol. 113, 1966, pp. 1823–1834.

23. "Recording of System Outage and Interruption Data," Electrical System and Equipment Committee Report, Edison Electric Institute, 1961.

24. Edison Electric Institute, "Report on Equipment Availability for the Ten Year Period 1960–1969," EEI Publication 70-26.

25. D. T. Egly and W. Chance, "Interruption Reporting," IEEE Paper, 33 PP 66-457.

3 APPLICATION TO GENERATION PLANNING

3.1 Introduction

This chapter discusses some aspects of the application of reliability calculation techniques to generating capacity planning. The material in this chapter is based mainly on five references.[1-5] As in all areas of reliability assessment it is necessary to consider the nature and usefulness of any reliability indices, efficient techniques for computing them, and the availability of system-component data for use in the models.

Most of the effort in the analysis of power-system reliability has been expended in the generation reserve requirement planning area.[6-8] A renewed interest in this particular problem naturally arises when efforts are made to integrate this work with the more recent efforts that have been devoted to reliability evaluation of transmission and distribution systems.[9-13] It becomes immediately obvious that there is a considerable divergence in the reliability measures commonly used in the various portions of the system and that they are apparently incompatible.

In the study of capacity-oriented generation reserve requirements, two reliability techniques are found in widespread use. The loss-of-load method is basically a calculation of the probability of the failure to be able to serve the expected peak loads over a specified time period. This technique includes the nature of the expected load and usually characterizes each individual generation unit by a maximum capability and a long-run probability of being in service, that is, its availability. The other technique applied is the frequency and duration technique presented by Halperin and Adler.[6] This method utilizes more data about each generation unit in that the average time durations of available and not-available, or repair, periods are used as well as the unit availability. The method allows the computation of the long-term probability of the generation system suffering an outage state of exactly a given amount and the expected frequency with which this state will recur. Reference 7 discusses these two methods in somewhat greater detail and gives a comparison of the numerical results obtainable in each case. There are also applications of probability methods in the assessment of the reliability of energy sources and the expected power-system production costs.[14,15]

Generation System Model

Increasing attention is being given to the development of reliability measures that are common to the entire system. At the distribution end of the system, reliability assessments and statistical performance records are apt to be kept on the basis of the number of outages, their frequency, their duration, and the number of customers affected. Methods of predicting these measures as an aid to more reliable transmission and distribution design have been developed.[9-13] It would seem highly desirable therefore to reexamine the techniques used for the generation system to develop a technique compatible with these methods.

This chapter sets up a model of the generation system that may be used for generation reserve studies and that may be used to compute availability, frequency, and duration for both exact and cumulative capacity outage states. It offers the opportunity for the eventual integration of the power generating system model into an overall power-system reliability model.

3.2 Generation System Model

In modeling the generation system, the units are assumed to be connected in parallel. Each unit is defined by a given maximum capability and by a long-run behavior pattern with regard to the occurrence of the available-repair cycles through which it passes. Each unit, in turn, may be merged into the generation system to permit the development of a capacity model characterized by the existence of various amounts of capacity available (or, conversely, on outage), the expected availability of exactly this capacity, and the expected recurrence, or cycle time, of this state.

The technique to be presented differs from previous techniques since in this model each unit may be described by its own capability and durations of available and repair periods, and the system model with exact capability states is readily transformed into one characterized by occurrences of a given amount of capacity-available or more. These are the same as the cumulative outage states in the loss-of-load probability model. As shown later, this transformation may be accomplished without any restrictive, unnecessary, or simplifying assumptions. The model developed may be used to implement both of the commonly used generation-system reliability computation techniques. It will generate information concerning the frequency and duration of cumulative outage

Application to Generation Planning 48

states, and it provides frequency and duration measures of the generation system compatible with the transmission and distribution system reliability measures.

The basic goals of the remainder of this chapter are to (1) present the logical development of a reliability model of a power generation system, (2) illustrate how recursive relationships may be derived and used to develop numerical data, and (3) compare the results of this method with other techniques by means of examples.

3.2.1 Generation System Model and Data Requirements

There are certain concepts that will be of value as this method is developed. The ability of a generator to provide power is equal to its instantaneous capacity. This is a value changing in time and dependent upon the state of the auxiliary equipment associated with the generator and the environment about the plant, such as the temperature of the condenser cooling water. The capacity may be at full machine rating for certain periods of time, changing suddenly to a partial rating due to the loss of certain auxiliary equipment, or it may be at zero value when the unit is taken out of service. The transitions from one capacity state to another are assumed to take place instantaneously and to occur at any time. The amount of time that the capacity remains at a certain value is the time in residence for the given state. The availability of a given state is then the mean time in residence divided by the mean cycle of time for this particular state to occur or recur.

Figure 2.1 shows a sketch of this phenomenon in which the occurrences of only one outage state are presented. It should be emphasized at the outset that it is not necessary to represent a generator simply as a binary machine with the capability of full output or none. One or more partial-capacity output states can be included if required. The theory to be presented is applicable to the binary, ternary, or multiple-state system. The examples selected are for binary machines.

3.2.2 Stochastic Processes, Markov Processes, and Transitions

A single repairable device that is either available (up) or in repair (i.e., down) may be characterized by its expected or mean behavior. It is assumed that repair and failure rates are constant. It is further assumed that the mean time-to-failure m and mean time-to-repair r are finite. The assumption of constant failure and repair rates puts the

Generation System Model

machine state description into the more restricted class of Markov processes. With finite r and m, both up and down states are "accessible," and over a long interval of time, the availability, or fraction of time the machine will be in an up state, is a number greater than zero and less than one.

The mean cycle depicted in Figure 2.1 defines the following terms:

$T = 1/f$, cycle time (days),
$f =$ frequency (cycle per unit time),
$m = 1/\lambda$, mean uptime (days),
and
$r = 1/\mu$, mean repair time (days).

In addition, these parameters may be used to define mean rates of occurrence and long-term availabilities as follows:

$\lambda =$ failure rate (failures per unit time),
$\mu =$ repair rate (repairs per unit time),
$A = m/(m + r) = m/T$, availability (steady state),
and
$\bar{A} = 1 - A = r/T$, unavailability (steady state).

The term "availability" is used to indicate the steady-state or long-time average availability. The availabilities, transition rates, and mean cycle time are related by

$$\lambda = \frac{1}{AT}, \qquad (3.1)$$

$$\mu = \frac{1}{\bar{A}T}, \qquad (3.2)$$

and

$$f = A\lambda = \bar{A}\mu. \qquad (3.3)$$

Figure 2.2 indicates the state transition diagram drawn for the two-state device. The arrows indicate entries or exits from the states, and the quantities $\lambda = 1/m$ and $\mu = 1/r$ designate the rates of departure

Application to Generation Planning

and entry. It should be reasonably self-evident that the frequency with which a state is encountered in the long run is as follows:

$$\begin{aligned} f_{\text{up}} &= A\lambda \\ &= \text{(steady-state probability of being in state)} \\ &\quad \times \text{(rate of departure)} \end{aligned} \quad (3.4)$$

or

$$\begin{aligned} f_{\text{up}} &= \bar{A}\mu \\ &= \text{(steady-state probability of not being in the state)} \\ &\quad \times \text{(rate of entry)}. \end{aligned} \quad (3.5)$$

EXAMPLE 3.1 A SINGLE, REPAIRABLE GENERATOR UNIT
Capacity = 20 MW,
$A = 0.98$,
$r = 2.040816$ days.

Therefore

$$\mu = \frac{1}{r} = 0.4900 \text{ per day,}$$

$$\lambda = \frac{\bar{A}}{rA} = 0.0100 \text{ per day,}$$

so that

$$T = \frac{r}{\bar{A}} = 102.0408 \text{ days}$$

is the mean cycle time for encountering either the up or down states.

3.2.3 Two Machines in Parallel

Equations 3.4 and 3.5 are perfectly general even if there is more than one mode of entering or leaving a state. The case of two repairable machines in parallel may be used to illustrate this. The number of possible states is $2^2 = 4$. The definitions of the four possible states are given in Table 3.1. Figure 3.1 shows the transition diagram for these states.

Generation System Model

Table 3.1 Possible States of Two Repairable Machines in Parallel

State Number	Machine 1	Machine 2	Rate of Departure
1	up	up	$\lambda_1 + \lambda_2$
2	down	up	$\mu_1 + \lambda_2$
3	up	down	$\lambda_1 + \mu_2$
4	down	down	$\mu_1 + \mu_2$

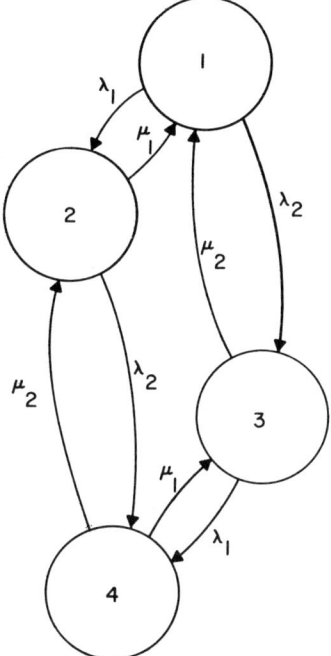

Figure 3.1 Four-state transition diagram for two repairable machines in parallel.

Application to Generation Planning

For this model, the run-repair process for each machine is independent of the process for the other machine. The last column in Table 3.1 indicates the rates of departure from each of the states. Note that the mean time in residence in a state is equal to the reciprocal of the rate of departure. As an example, the cycle time between encountering state 2, on the average, is

$$T_2 = \frac{1}{A_{\text{state } 2}(\lambda_2 + \mu_1)}.$$

EXAMPLE 3.2 TWO GENERATORS IN PARALLEL
See Table 3.2.

Referring to the state transition diagram of Figure 3.1, the availabilities and mean times between encountering the states are given in Table 3.3.

3.2.4 Cumulative-Event Cycle Time

In the loss-of-load probability method, it is convenient to redefine the events considered so that an outage "event" is the occurrence of an

Table 3.2 Description of the Two Generators of Example 3.2

Unit	Capacity (MW)	Availability	r (days)	μ (per day)	λ (per day)
1	20	0.9800	2.040816	0.49	0.01
2	30	0.9800	2.040816	0.49	0.01

Table 3.3 Description of States for Example 3.2

State Number	Capacity Available (MW)	A (per unit)	Rate of Departure (per day)	Cycle Time (days)
1	50	0.9604	$\lambda_1 + \lambda_2 = 0.02$	52.0616
2	30	0.0196	$\mu_1 + \lambda_2 = 0.50$	102.0408
3	20	0.0196	$\lambda_1 + \mu_2 = 0.50$	102.0408
4	0	0.0004	$\mu_1 + \mu_2 = 0.98$	2551.02

Generation System Model 53

outage of a given magnitude or greater. The same reasoning holds true in the frequency-duration method. It is interesting to note that the mean time between encountering an outage of exactly 30 MW is 102.0408 days, but it would be of more value to know the frequency of encountering an outage of 30 MW or more. That is, how often (frequency) will the outage change from a value less than 30 MW to an outage of 30 MW or more?

To accomplish this, it is merely necessary to redefine the state so that each state is now the occurrence of a given capacity outage or larger. The previous, two-parallel-machine transition diagram may be used to illustrate this transformation procedure and the steps necessary to obtain the frequency of encountering the newly defined states. Figure 3.2 shows the new states superimposed over the old. The new, or cumulative, states are denoted by primes. Notice that the new states are

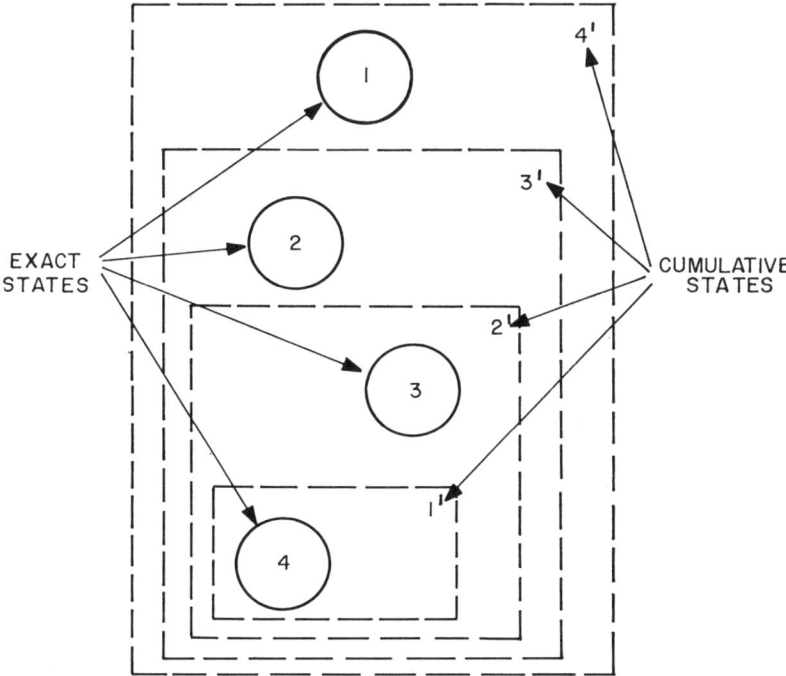

Figure 3.2 Relationship of exact and cumulative state descriptions for the two-machine example.

Application to Generation Planning

numbered differently from the old: state $1'$ = state 4, state $2'$ = states 3 and 4, and so on.

The frequency of encountering state $1'$ is the same as that of encountering state 4:

$$f_{1'} = A_4(\mu_1 + \mu_2) = f_4.$$

The frequency of encountering the new state $2'$ is equal to the sum of the frequencies with which transfers take place from old state 3 to old state 1, $A_3\mu_2$, and from old state 4 to old state 2, $A_4\mu_2$. This result is less than the sum of the frequencies of encountering state 3 or 4 by the sum of the encounters of state 4 from state 3, $A_3\lambda_1$, and of state 3 from state 4, $A_4\mu_1$. Note that transfers between states 3 and 4 represent failure and repair of machine 1. The frequency of transfer either way is given by the product of the unavailability of machine 2 and the frequency of encounter of machine 1 "up" state. To illustrate, the frequency of encounter of cumulative state $2', f_{2'}$, is given by the sum of the frequency of encounter of state $1', f_{1'}$, and the frequency of encounter of old state 3 from old state 1, $A_3\mu_2$, less the frequency of encounter of old state 4 from old state 3, $A_3\lambda_1$:

$$f_{2'} = f_{1'} - A_3\lambda_1 + A_3\mu_2.$$

The following transition rates are defined so that the example may be generalized.

Let

$\lambda_{+k} = \lambda_{\text{up}}$ = rate of transition out of a given capacity state k to one in which more capacity is available

and

$\lambda_{-k} = \lambda_{\text{down}}$ = rate of transition out of a given capacity state k to one in which less capacity is available.

The frequency of encountering a state with a given capacity or less is given by the recursive relationship in Equation 3.6. In this relationship

Generation System Model

exact (i.e., unprimed) state k is being added to cumulative state $n-1$ (i.e., primed) to obtain the new cumulative capacity state n:

$$f_n = f_{n-1} - A_k \lambda_{-k} + A_k \lambda_{+k}. \tag{3.6}$$

In Equation 3.6, A_k is the availability of the exact capacity state k, and the primes have been discarded and replaced by the subscripts n and $n-1$.

EXAMPLE 3.3 TWO GENERATORS IN PARALLEL
This example continues Example 3.2 to calculate the availability and cycle times of the cumulative capacity states. For the sake of completeness, the availability of a cumulative capacity state n may also be found from the relationship

$$A_n = A_{n-1} + A_k, \tag{3.7}$$

where, again, the exact capacity state k is being appended to the cumulative state $n-1$ to arrive at n.

The necessary data have been generated in the first two examples. The transition diagram for the two machines in parallel plus these data permit the calculation of λ_{up}, λ_{down}, A_n, and f_n for the four cumulative capacity states. Table 3.4 gives these results but with the cycle time instead of the frequency.

3.2.5 Identical Capacity States
In the construction of the exact capacity-state availability and frequency tables for larger systems, identical capacity states may be generated by different combinations of units. In the sense of the transition diagrams, there is no direct linkage between these states. The only way that a system may transit within a given instant of time from one exact capacity state to another state with the same capacity available is to have one machine repaired and another fail within the same instant. The probability of this occurring is of second order. It is so unlikely that it can be ignored relative to the occurrence of a single event. The two capacity states may therefore be merged as states separated in time.

Since transfer cannot occur directly from one state to the other, their availabilities and frequencies of encountering will add directly.

Application to Generation Planning

Table 3.4 Two-Machine Example

Exact Capacity States			Departure Rates	
State Number	Capacity (MW)	Availability	λ_{up} (per day)	λ_{down} (per day)
1	50	0.9604	0	0.02
2	30	0.0196	0.490	0.01
3	20	0.0196	0.490	0.01
4	0	0.0004	0.980	0

Cumulative Capacity States			
State Number	Capacity (MW)	Availability	Cycle Time (days)
4	50	1.0000	
3	30	0.0396	52.0616
2	20	0.0200	102.0408
1	0	0.0004	2551.02

Let j and i designate two states that have exactly the same capacity available and k designate the merged state. The capacity, availability, and cycle frequency of the merged state are as follows:

Capacity:

$$C_k = C_i = C_j. \tag{3.8}$$

Availability:

$$A_k = A_i + A_j. \tag{3.9}$$

Frequency:

$$f_k = f_i + f_j. \tag{3.10}$$

The total rates of departure to greater and lesser capacity states may be found from

$$A_k \lambda_{up,k} = A_i \lambda_{up,i} + A_j \lambda_{up,j} \tag{3.11}$$

and

$$A_k \lambda_{down,k} = A_i \lambda_{down,i} + A_j \lambda_{down,j}. \tag{3.12}$$

Generation System Model 57

These relationships complete the set required to permit construction of nonredundant, exact capacity-availability tables. With the previous developments, cumulative capacity data may then be generated recursively. This model has been implemented in several digital computer programs, and the next section illustrates some numerical results for larger systems.

3.2.6 Numerical Examples

Since the practical applications of probabilistic models of generation systems tend to make more use of the data for the cumulative capacity-available (or outage) states, the numerical results will emphasize these data. The first data are presented to illustrate the type of information that may be obtained for a collection of dissimilar units without the necessity of making any simplifying approximations. The second example verifies the recursive method developed here by comparison with mean times to failure and repair obtained for a group of identical machines. These later results were obtained by application of formulas developed as an outcome of an investigation of general system reliability.[16] Finally, the last example is a 22-machine case. The data for this system are used to illustrate the results obtained by the use of the exact formulation, and these are compared with the results of methods that have been previously suggested.

EXAMPLE 3.4 A FIVE-MACHINE SYSTEM

The system for this example is composed of five machines with the characteristics shown in Table 3.5.

The results of using the relationship developed earlier are given in

Table 3.5 Characteristics of the Machines in Example 3.4

Capacity (MW)	Mean Repair Time (days)	Availability (per unit)
20	2.040816	0.980
30	2.040816	0.980
40	5.1082	0.975
50	5.1543	0.970
60	5.1543	0.970
Total 200		

Application to Generation Planning

Table 3.6. These data include the cycle times and the availabilities for both the exact and cumulative outage states.

Figure 3.3 illustrates the cycle time and frequency of recurrence of the cumulative outage states. These data along with the more well-known information about the existence probability (i.e., the availability) of the cumulative outage states provide a comprehensive reliability picture of the generation system. Note that the frequency of occurrence of a zero outage or more is zero, meaning that the system is always in this state. The data of Figure 3.3 are shown as smooth curves for the

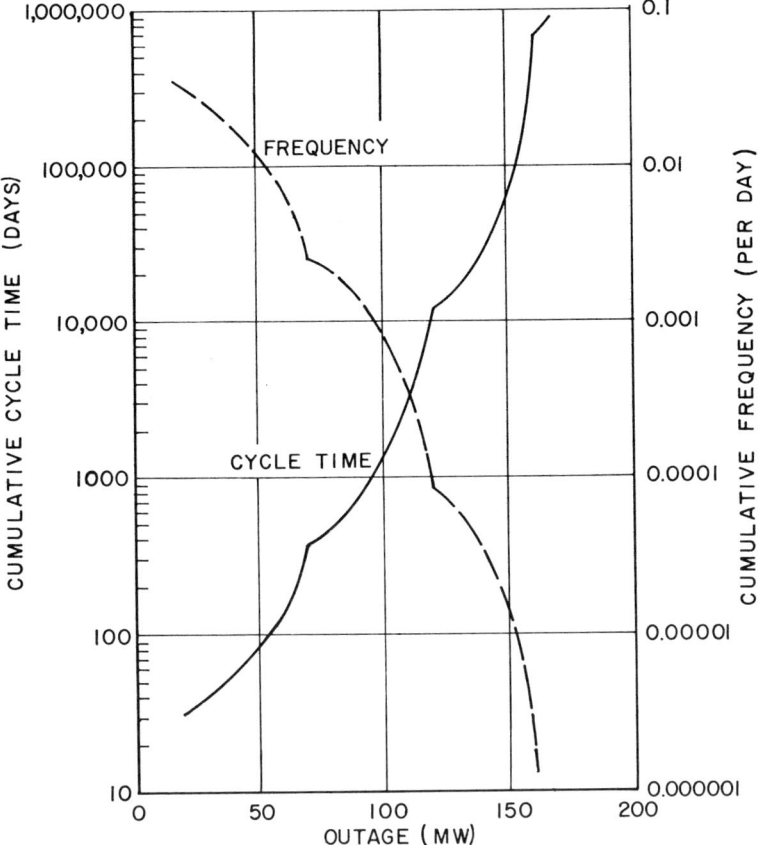

Figure 3.3 Cumulative state data for five-machine system. Cycle time is mean time to recurrence of an outage of a given magnitude or more. Frequency is reciprocal of recurrence time.

Table 3.6 Five-Machine Example

Capacity Outage (MW)	Exact Outage States		Cumulative Outage States	
	Availability	Cycle Time (days)	Availability	Cycle Time (days)
0	0.88105	30.659	1.00000	—
20	0.01798	107.57	0.11895	30.659
30	0.01798	107.57	0.10097	41.166
40	0.02259	194.35	0.08299	62.627
50	0.02762	153.90	0.06040	81.514
60	0.02771	154.84	0.03278	133.81
70	0.10171×10^{-2}	1392.0	0.50726×10^{-2}	366.87
80	0.11122×10^{-2}	1275.3	0.40554×10^{-2}	487.56
90	0.12642×10^{-2}	1441.4	0.29432×10^{-2}	761.24
100	0.71004×10^{-3}	3290.0	0.16790×10^{-2}	1470.9
110	0.86836×10^{-3}	2671.5	0.96897×10^{-3}	2424.4
120	0.28518×10^{-4}	3.9145×10^4	0.10061×10^{-3}	1.2339×10^4
130	0.31458×10^{-4}	3.5546×10^4	0.72090×10^{-4}	1.7728×10^4
140	0.17490×10^{-4}	6.3452×10^4	0.40632×10^{-4}	3.4188×10^4
150	0.21900×10^{-4}	7.4362×10^4	0.23142×10^{-4}	7.1375×10^4

Table 3:7 20 Identical Machines

Number in Service	Availability	Cumulative State Results Cycle Time (days)	Mean Time-to-Failure—MTTF (days)	Mean Time-to-Repair—MTTR (days)
20	1.0	—	—	—
19	0.33239	18.349	12.250	6.099
18	0.05990	47.321	44.487	2.834
17	0.70687×10^{-2}	257.64	255.82	1.821
16	0.59968×10^{-2}	2227.82	2226.48	1.336
15	0.38591×10^{-4}	2.7291×10^4	2.7290×10^4	1.053
14	0.19484×10^{-5}	4.4575×10^5	4.4575×10^5	0.868
13	0.78908×10^{-7}	9.3607×10^6	9.3607×10^6	0.739
12	0.26010×10^{-8}	2.4698×10^8	2.4698×10^8	0.642
11	0.70434×10^{-10}	8.0680×10^9	8.0680×10^9	0.568

Generation System Model 61

sake of clarity. Actually, they should be stepped to represent the changes that take place as specific units change their status.

EXAMPLE 3.5 A SYSTEM OF 20 IDENTICAL UNITS
Einhorn, in a general study of the reliability of a system of binary-state, repairable elements, each with independent, exponentially distributed times in service and times on repair, has developed relations for the reliability of systems of paralleled units.[16] Specifically, the relationships for the availability, mean time-to-failure (MTTF), and mean time-to-repair (MTTR) of r units out of a total of n may be utilized to verify the recursive relationships. Einhorn's analysis assumes constant failure and repair rates and develops the expected (or mean) values of various system-reliability measures. Einhorn's results form the basis for checking the relationships derived here.

The example assumes 20 identical machines each of which has a mean repair time of 5 days and an unavailability (i.e., forced outage rate) of 0.02. The computations were made in two ways. Table 3.7 gives the numerical results obtained using Einhorn's equations. The sum of the mean time-to-failure plus the mean time-to-repair is the mean cycle time. For example, for the 16-machine cumulative state, the MTTF is 2226.48 days and the MTTR is 1.34 days, resulting in a mean cycle time of 2227.82 days.

This same division of the total cycle time between mean times to repair and failure have been computed using the methods of this chapter by making appropriate use of Equations 3.4 and 3.5 along with the availabilities and rates of transfer up and down out of the cumulative states. The total cycle times and availabilities of the cumulative capacity states were computed with the recursive equations and found to check exactly the results obtained with Einhorn's method.

As may be seen from the data of Table 3.7, the MTTF is very much larger than the MTTR for all the states except those where only a few units are out of service. Thus, in all practical cases, the total cycle time effectively measures the MTTF.

EXAMPLE 3.6 A 22-UNIT MODEL SYSTEM
A system that has been used in previous publications to illustrate various techniques is the system of 22 units suggested by Arnoff and

Table 3.8 Characteristics of the Machines in Example 3.6

Number of Identical Units	Unit Size (MW)	Mean Repair Time r (yr)	Mean Cycle Time T (yr)
1	250	0.06	3.0
3	150	0.06	3.0
2	100	0.06	3.0
4	75	0.06	3.0
9	50	0.06	3.0
3	25	0.06	3.0
Total 22	1725		

Chambers.[17,18] The relevant data for the units in this system are listed in Table 3.8.

Table 3.9 gives the results of applying different techniques to calculating the frequencies of the exact and cumulative states.

The Halperin-Adler method and the method of this chapter were used to compute the frequency of occurrence of the exact outage states. These are the data in the third and fourth column of Table 3.9. The deviations between the two methods rarely go beyond 20%, indicating the applicability of the technique of Halperin and Adler[6] for obtaining frequency and duration data for the exact outage states.

The calculation of the cumulative frequency has been done in three ways. The fourth and fifth columns of Table 3.9 compare the results obtained by Sauter, Baldwin, and Dale[17] with those obtained by the recursive technique. Sauter and his co-workers suggested a technique that effectively ignored the transfer of the system within a given cumulative outage state. They suggested in effect that the recursive development of the frequency of the cumulative state be approximated by

$$f_n = f_{n-1} + f_k,$$

where f_k is the frequency of the exact state being added to the $(n-1)$th cumulative state to get the nth cumulative state. Halperin and Adler on the other hand, suggested a machine grouping technique for obtaining

Table 3.9 Results for 22-Machine System

Outage (MW)	Availability of Exact Outage	Frequency of Exact Outage		Frequency of Outage or More	
		Halperin and Adler [6] (per yr)	Recursive (per yr)	Sauter et al. [17] (per yr)	Recursive (per yr)
0	0.641171	4.797978	4.79788	—	—
25	0.039255	0.934665	0.934651	8.829087	4.79788
50	0.118567	2.754298	2.83611	7.894422	4.42402
75	0.059556	1.495380	1.53590	5.140124	3.28128
100	0.039136	1.126330	1.14588	3.644744	2.59117
125	0.011871	0.540627	0.487157	2.518414	1.99545
150	0.046756	1.259236	1.25372	1.977787	1.66536
175	0.005757	0.252479	0.251611	1.718551	1.07773
200	0.008414	0.365578	0.354191	1.466072	0.90359
225	0.004161	0.185825	0.184019	1.100494	0.66320
250	0.015668	0.432936	0.432576	0.914669	0.53508
275	0.001597	0.083256	0.078599	0.481733	0.32384
300	0.003702	0.177995	0.158219	0.398477	0.26637
325	0.001478	0.067382	0.067440	0.220482	0.15811
350	0.001020	0.049107	0.049960	0.153100	0.11044
375	0.000349	0.021984	0.020688	0.103993	0.07398
400	0.001015	0.045499	0.045120	0.082009	0.05776
425	0.000137	0.011211	0.008512	0.036510	0.02626

Table 3.9 Results for 22-Machine System (continued)

Outage (MW)	Availability of Exact Outage	Frequency of Exact Outage		Frequency of Outage or More	
		Halperin and Adler [6] (per yr)	Recursive (per yr)	Sauter et al. [17] (per yr)	Recursive (per yr)
450	0.000188	0.012059	0.011169	0.025299	0.01948
475	0.000090	0.005470	0.005544	0.013240	0.01072
500	0.000055	0.003467	0.003543	0.007778	0.00632
525	0.000017	0.001346	0.001315	0.004311	0.00347
550	0.000027	0.001921	0.001718	0.002965	0.00237
575	0.000006	0.000408	0.000437	0.001044	0.00099
600	0.000005	0.000382	0.000359	0.000626	0.00062
625	0.000002	0.000162	0.000178	0.000244	0.00031
650	0.000001	0.000082	0.000102	0.000082	0.00016

Generation System Model

a smooth curve estimating cumulative outage frequency as a function of the outage.[6] Their method was used with a 100-MW group size for this system.

The results of applying all three methods are shown in Figure 3.4. Three curves are shown: (1) the curve obtained using the recursive formulation, (2) the "100-MW grouping" obtained by using the method suggested by Halperin and Adler,[6] and (3) the curve obtained using the technique of Sauter, Baldwin, and Dale.[17] These results indicate the inherent error in using either approximation to the cumulative frequency relationship.

3.2.6.1 Comments The last three examples illustrate the results obtainable

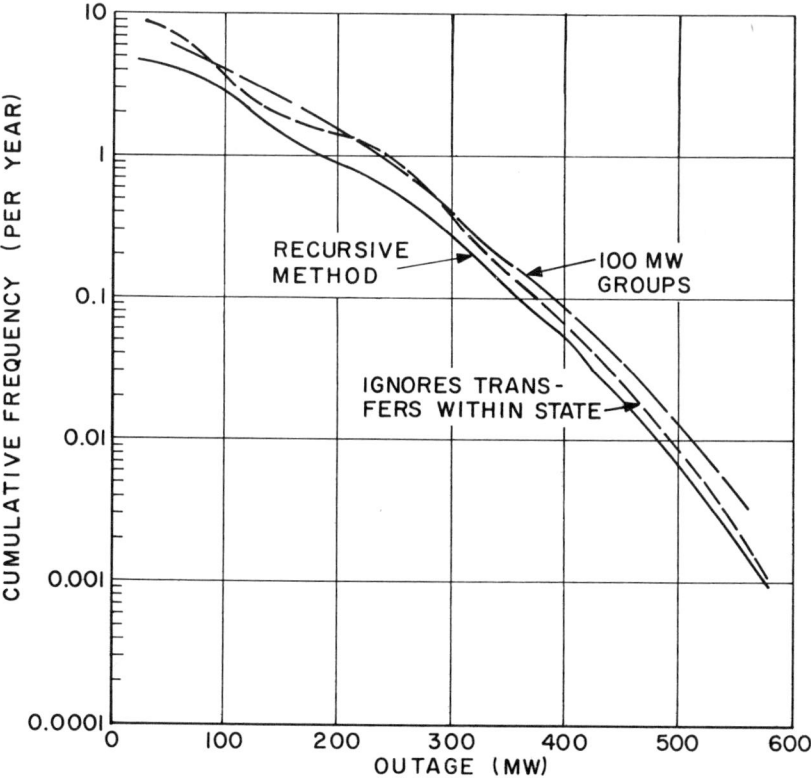

Figure 3.4 Cumulative frequency curves for the 22-unit system. The exact data are compared with two approximations.

by modeling the generation system by a frequency and duration method. They substantiate the validity of the equations developed in the previous section, and they point out the effect of approximations on the calculations for the cumulative capacity states. The present modeling method develops the capacity outage and availability data as well as the cycle times. It may also be used to generate information concerning the mean duration of both exact and cumulative outage states. However, as would be expected and as shown by the results of Example 3.5, the mean repair times are very much less than the mean uptimes (i.e., MTTF). Perhaps the most significant result of all of these examples is the demonstration that frequency and duration of outages may be generated recursively in a quite straightforward fashion.

3.2.7 Data Requirements

The foregoing analysis is only as useful as the availability of well-processed data on equipment reliability will permit. It has been a long and continuing effort to produce equipment-outage-data statistics in the power-system field. Both the IEEE and industry groups such as the Edison Electric Institute have taken the lead in collecting and processing data and making them available to the industry. For a number of years these data were available only in the form of annually published reports put out by the EEI.[19] In recent years new definitions of terms and data specifications have been evolved by industry groups. These plus the availability of equipment outage records on magnetic tape have facilitated the collection, processing, and use of equipment outage data.

The collection of historical data does not circumvent a fundamental problem in power-system planning: forecasting the outage time of new equipments of larger sizes and perhaps different designs than those used in the past. The historical records can, however, give indications of what may be expected in the future. For instance, the introduction of new types of generation and the extension of design to new size ranges seem to be accompanied by a transient increase in forced outage rates that may be damped out after several years of experience with a new design. The problem of forecasting the expected reliability of future system equipment elements is one that the planner cannot escape. It is not possible to foresee the future, but it is absolutely necessary to learn from the past.

3.3 Load Model

The results of this description of the generation-capacity model are useful in themselves as measures of the reliability of the generation system, but it would seem that a more adequate measure would be one that incorporated a consideration of the expected load pattern. The reliability model used for planning generation systems should incorporate calculations of both the availability (i.e., probability of existence) and the occurrence frequency of cumulative capacity-reserve-margin states. The system model should result in predictions describing the conditions of generating capacity in excess of the load by a given amount or more. This, of course, requires that the previously developed capacity model be combined with (i.e., "convolved") a suitable load model. A load model is suggested and data describing the load states, state availabilities, and frequencies of occurrence of the various states are given. The model permits the consideration of the fact that daily peak loads usually do not persist for a 24-hour period. The load model is based on a Markov chain.

The load model may then be merged with the generating-capacity model to permit the calculation of system reserve and reliability measures. This is done assuming that the load occurrence and capacity reliability processes are statistically independent. The recursive procedure suggested has proved to be very efficient computationally. These developments are illustrated by combining the two-machine, 50-MW system with the load model describing the loads for a 20-day interval.

The sequence of daily peak loads is assumed to be a stationary, random process.[2] This model seems to be suitable for the problems associated with long-range generating-capacity planning. It is compatible with the load model used in the loss-of-load method and is amenable to analytic treatment. It also provides a means for extending the usual loss-of-capacity technique to incorporate a consideration of load distributions. The model represents the daily load cycle as a sequence of peak loads L_i, each of a mean duration of e day interspersed with periods averaging $(1 - e)$ day of a fixed, light load L_0. As illustrated in Figure 3.5 the sequence of peak loads is random.

3.3.1 Load Model for Planning

The load model is based on the following:

Application to Generation Planning 68

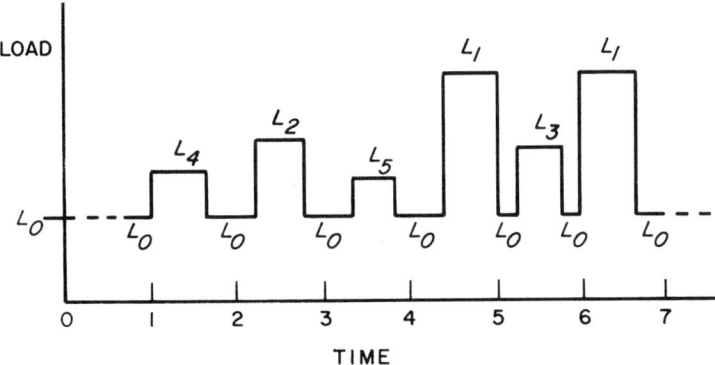

Figure 3.5 Sequence of loads for basic load model.

1. Daily loads in a period will be represented by a set of N load levels or load states.
2. The sequence of daily peak loads is a random sequence of the N load states.
3. The load model is statistically stationary.
4. The distribution of residence times in a given load state is exponential.
5. At each transfer to a new load state the probability of transfer to a particular state is directly proportional to the long-term, average (i.e., quiescent or steady-state) probability of existence of the new state (i.e., its "availability") A_i.
6. Load state transitions occur independently of generation state transitions.
7. The mean duration of peak loads is the fraction e day, with each succession of peak loads separated by a low-load period of $(1 - e)$ day mean duration.

Reference 2 discusses some of the mathematical aspects of the load model when it is considered as a Markov chain. Table 3.10 gives a summary of the results of the mathematical analysis of the basic model.

These seven assumptions allow the construction of a stochastic load model that may be merged with the capacity-availability model to produce an overall system description in terms of reserve margin states, their probability of existence (i.e., availability), and the expected time between recurrences.

The same general procedure may be followed in developing the load model as that used in analyzing the various capacity states. We need

Load Model

Table 3.10 Summary of Load Model

Number of load levels	N
Description of load level, MW	$L_i, i = 1, 2, \ldots, N$
Number of occurrences of L_i	$n_i, i = 1, \ldots, N$
Interval length, days	$D = \sum_{i=1}^{N} n_i$
Expected duration of peak, days	$e < 1$
"Availability" of L_i, p.u.	$A_i = n_i e / D$
Transition rate to greater load	$\lambda_{+L_i} = 0$
Transition rate to lesser load	$\lambda_{-L_i} = 1/e$
Frequency of occurrence of L_i	$f_i = n_i / D$
For Low Load Period:	
Load state	L_0, MW
Availability	$A_0 = 1 - e$
Transition rates	$\lambda_{-L_0} = 0$
	$\lambda_{+L_0} = \lambda_0 = 1/(1 - e)$
Frequency	$f_0 = 1$

relationships establishing the availability of each load level, the cycle time or frequency of each exact load level, the rates of departure from each state, and the availability and frequency of encountering a given load level or less. An alternative load model that assumes direct transfers between peak load states is developed in Reference 2.

3.3.2 Annual Quantities

To represent a year for either the generation or load, some means must be found to recognize the nonstationary effects of scheduled maintenance and seasonal load changes. A reasonable approach that has been frequently used in the past is to divide the year into "maintenance" intervals. These intervals, say 4 weeks' duration or less, are short enough so that within the interval the lists of machines on scheduled maintenance may be reasonably assumed not to change. This period is also short enough so that the load model may be represented by a stationary stochastic process.

To relate the interval representation to a year, the values of A_i must be multiplied by the fraction of the year represented by the interval, $D/365$. The mean duration of residence of a given load level will not

Application to Generation Planning

change, nor will there be a change in the transition rates to higher or lower loads from a given load level.

Each peak load lasts e fraction of a day on the average and is always followed by a low-load period. The availability of the particular peak load level L_i in the interval D days long is given by

$$A_i = \frac{en_i}{D}, \tag{3.13}$$

where n_i is the expected number of occurrences of L_i in the interval of D days. Therefore if one relates the interval representation to a yearly base of 365 days, the availability of the load state L_i becomes $en_i/365$.

Regardless of interval or yearly representation, the mean duration at load level L_i is e day. Since by definition, the model precludes transfer from one peak-load state to another, the probability, given the load is in state L_i, that the next transfer is to a low-load state is 1, that is, a certainty. Accordingly, the transition rate to a lower load state is $1/e$ and to a higher load state, zero.

3.3.3 Load Model Results

To illustrate the procedure and form of the basic load model, assume that during a 20-day period, peak loads of 40, 25, 20, and 15 MW are expected to occur 2, 5, 8, and 5 days, respectively. Let the duration e of each peak load be 0.5 day. Results are given in Table 3.11.

Data for the cumulative load states could be developed in a manner similar to that for the capacity model. Instead, the state availability and frequency of recurrence of various reserve, or margin, states are developed next.

3.3.3.1 Reserve Margin States Reserve, or margin, is the difference between the available capacity and the load. The capacity data and the load statistics developed earlier may be combined to develop data on the availability and frequency of recurrence of various degrees of reserve condition. A margin state m_k is the combination of the load state L_i and the capacity state C_j:

$$m_k = C_j - L_i. \tag{3.14}$$

Load Model

Table 3.11 Load Model Data

State Number	Load L_i (MW)	Occurrence n_i (days)	Availability A_i ($en_i/365$)	Cycle Time T_i (days)	Departure Rates (per day) λ_{up}	λ_{down}
1	40	2	0.00273973	182.5	0	2
2	25	5	0.00684932	73.0	0	2
3	20	8	0.01095890	45.625	0	2
4	15	5	0.00684932	73.0	0	2
0	0	20	0.02739727	18.25	2	0

$e = 0.5$ day

Application to Generation Planning

In order to calculate data on a cumulative margin basis it is necessary to calculate the rates of departure from m_k to larger and smaller margin states. Let subscripts $+$ and $-$ designate up and down transitions, and the subscripts m, c, and L designate margin, capacity, and load states. Then

$$\lambda_{+m} = \lambda_{+c} + \lambda_{-L}$$

and

$$\lambda_{-m} = \lambda_{-c} + \lambda_{+L}.$$

Given the independent load and capacity states, the rate of transfer from a given margin state to one with larger margin is equal to the rate of transfer upward in capacity plus the rate of transfer downward in load.

The data from the previous example may be used to illustrate the construction of a margin-availability table. The capacity data refer to Example 3.2 in which the two-machine system represented consists of 20- and 30-MW units.

3.3.3.2 Margin-Availablity Tables Table 3.12 illustrates the construction of margin-availability tables for the "exact" margin states that occur in the two-machine, four-load-level example. The data in the table include margin in megawatts, availability, and transfer rates. The states shown include all combinations of load and capacity. As a result there are entries in the table that represent identical margin values but that are the result of different combinations of loads and capacities. The data are for a 365-day year and assume the exposure factor e is 0.5. Entries for the zero, or low-load, level are not shown. In the generation of these exact states the availability of the margin state is

$$A_k = A_j A_i, \tag{3.15}$$

and the departure rates are given by

$$\lambda_{\pm k} = \lambda_{\pm j} + \lambda_{\mp i}. \tag{3.16}$$

Load Model

Table 3.12 Margin States

Generation Data					Load Data				
j	C_j	A_j	λ_+	λ_-	$i =$	1	2	3	4
					$L_i =$	40	25	20	15
					$A_i =$	0.00273973	0.00684932	0.0109589	0.00684932
					$\lambda_+ =$	0	0	0	0
					$\lambda_- =$	2	2	2	2
1	50	0.9604	0	0.02	$m =$	10	25	30	35
					$A =$	0.002631	0.006578	0.010525	0.006578
					$\lambda_+ =$	2	2	2	2
					$\lambda_- =$	0.02	0.02	0.02	0.02
2	30	0.0196	0.49	0.01	$m =$	-10	5	10	15
					$A =$	0.000054	0.000134	0.000215	0.000134
					$\lambda_+ =$	2.49	2.49	2.49	2.49
					$\lambda_- =$	0.01	0.01	0.01	0.01
3	20	0.0196	0.49	0.01	$m =$	-20	-5	0	5
					$A =$	0.000054	0.000134	0.000215	0.000134
					$\lambda_+ =$	2.49	2.49	2.49	2.49
					$\lambda_- =$	0.01	0.01	0.01	0.01
4	0	0.0004	0.98	0	$m =$	-40	-25	-20	-15
					$A =$	0.000001	0.000003	0.000004	0.000003
					$\lambda_+ =$	2.98	2.98	2.98	2.98
					$\lambda_- =$	0	0	0	0

$e = 0.5$ day

Application to Generation Planning

These may be combined to yield the occurrence frequency of these exact states by means of a modification of Equation 3.3:

$$f_k = A_k(\lambda_{+k} + \lambda_{-k}). \tag{3.17}$$

The margin and capacity states may be arrayed in tabular form prior to elimination of identical margin entries. The identical margin states may be combined as follows: For a given state m_k made up of N identical margin states such that $m_k = m_1 = m_2 = \cdots = m_N$,

$$A_k = \sum_{\ell=1}^{N} A_\ell, \tag{3.18}$$

$$f_k = \sum_{\ell=1}^{N} f_\ell, \tag{3.19}$$

and

$$\lambda_k = \sum_{\ell=1}^{N} \frac{A_\ell \lambda_\ell}{A_k}, \tag{3.20}$$

where the last relationship applies to either the upward or the downward departure rates. In the development of Equations 3.19 and 3.20 use was made of the observation that transfers cannot take place between two identical margin states without an intervening state of differing margin.

3.3.3.3 Cumulative Margin States As with the capacity data, it is highly desirable to have availabilities and frequencies for cumulative margin states. With distinct, ordered tables of data for the exact margin states available, the cumulative data may be developed exactly as was done for the cumulative capacity data. That is, Equations 3.6 and 3.7 may be used directly, with the understanding that they are to be applied to the reserve margin data.

It has been found that more computationally efficient techniques result if the data describing cumulative generation states are combined with information for the exact load states. These algorithms are developed now. In order to distinguish between exact and cumulative state descriptions, the following definitions are made:

Load Model

m exact margin of m, MW,
M margin of M, MW, or less,
L exact load of L, MW,
C exact capacity of C, MW,
and
G cumulative capacity of G, MW, or less.

For a given cumulative margin state the availability is

$$A_M = \sum_{m \leq M} A_m. \tag{3.21}$$

For any exact margin state, $m = C - L$, the availability is

$$A_m = \sum_{C,L} A_C A_L, \tag{3.22}$$

where the sum is over all events where the relation between C, L, and m is true. The cumulative margin state availability for a margin of M or less MW is

$$A_M = \sum_{m \leq M} \sum_{C,L} A_L A_C. \tag{3.23}$$

Since load and capacity states are independent, Equation 3.23 may be rearranged by summing over all load levels. That is,

$$A_M = \sum_L A_L \sum_{C \leq L+M} A_C. \tag{3.24}$$

This last sum is the availability of the cumulative capacity state G, defined by the set of events

$$\{G\} = \{C \leq L + M\}; \tag{3.25}$$

thus,

$$A_M = \sum_L A_L A_G. \tag{3.26}$$

For example, consider the four-state load and two-machine sample problem developed previously. Let $M = -10$ MW be the cumulative

Application to Generation Planning

margin state under consideration. The availability of that cumulative margin state is found from the availabilities of the following load and capacity states:

Margin of M: -10 MW or less.
Load: 40 25 20 15 MW exactly.
Capacity: 30 0 0 0 MW or less.

An efficient recursive relationship may also be developed for the frequency of occurrence of the cumulative margin states in terms of the exact load state conditions and the cumulative capacity data. The general relationship for the frequency of occurrence of cumulative states is given in Equation 3.6. By a minor modification of this relation, and in terms of exact margin states, the frequency of occurrence of a margin of M or less MW is

$$f_M = \sum_{m \leq M} A_m(\lambda_{+m} - \lambda_{-m}). \tag{3.27}$$

The definitions of the transition rates and the independence of the load and capacity events permit writing this as

$$f_M = \sum_{m \leq M} \sum_{L,C} A_L A_C (\lambda_{+C} - \lambda_{-C} + \lambda_{-L} - \lambda_{+L}), \tag{3.28}$$

where the exact load, capacity, and margin states are again related by $m = C - L$.

The same technique of summing over each exact load state may be used. That is,

$$f_M = \sum_L A_L \sum_{C \leq L+M} [A_C(\lambda_{+C} - \lambda_{-C}) + A_C(\lambda_{-L} - \lambda_{+L})]. \tag{3.29}$$

If use is made of the definition of the cumulative capacity state G, this relationship may be simplified since

$$f_G = \sum_{C \leq L+M} A_C(\lambda_{+C} - \lambda_{-C}) \tag{3.30}$$

is the frequency of the cumulative capacity state G. Therefore, the fre-

Load Model

quency of occurrence of the cumulative margin state is

$$_M = \sum_L A_L[f_G + A_G(\lambda_{-L} - \lambda_{+L})]. \tag{3.31}$$

These relationships have been found to be much more efficient computationally than combination of the exact margin states.

Equations 3.31 and 3.26 give frequency and availability, or probability of existence, of the cumulative margin states. That is, the availability number gives the probability of finding a specified cumulative margin state at any randomly selected time. This may be converted to the usual annual loss-of-load probability index by multiplying by the number of days in a year and dividing by the exposure factor e.

As an example, the frequency of occurrence of the cumulative margin state of -10 MW may be computed for the example using the load data given previously and the capacity data from

$$f_{-10} = A_{40}(f_{30} + A_{30}\lambda_{-L}) + (A_{25} + A_{20} + A_{15})(f_0 + A_0\lambda_{-L})$$

$$= 0.00273973 \left(\frac{1}{52.0616} + 0.0792\right)$$

$$+ 0.02465754 \left(\frac{1}{2551.2} + 0.0008\right)$$

$$= 0.000299 \text{ per day,}$$

which is the reciprocal of the cycle time given in Table 3.13.

At this point, it is reasonable to have reservations about the adequacy of the Markov model for representation of the sequence of daily peak loads. The times of residence in a given peak-load state L_i form an exponentially distributed random sequence with mean value equal to e. The times of residence in the low-load state L_0 form an exponentially distributed random sequence with mean value equal to $(1 - e)$. Only in the mean sense is the Markov model representative of the sequence of daily loads. The calculations for frequency of occurrence and for probability of existence of a given state are, by definition, mean values. Therefore, no more is asked from the Markov model than it can

Table 3.13 Cumulative Margin Data

Margin (MW)	Availability of Cumulative State (p.u.) $e = 0.5$ day	Cycle Time of Cumulative Events (days)
35	0.0273973	18.25
30	0.0208192	23.94
25	0.0102942	47.78
15	0.0037162	126.5
10	0.0035819	132.0
5	0.00073589	546.3
0	0.00046740	858.6
−5	0.00025260	1582.4
−10	0.00011836	3344.4
−15	0.000064658	6030.3
−20	0.000061918	6342.5
−25	0.000003836	87,488
−40	0.000001095	306,208

provide, nor, conversely does the model restrict the development of mean values. There are many stochastic models that would give the same probability and mean duration in a given state as the Markov model, but the Markov model happens to be an especially simple one to deal with.

3.3.4 Annual Quantities from Interval Data

The data of Table 3.13 apply for a single 20-day interval. The results are expressed in equivalent annual values since the load state availabilities were expressed on an annual basis. If there are data for the several consecutive intervals whose lengths sum up to the 365-day year, these interval data may be combined quite readily into annual values by utilizing the assumptions of statistical independence.

For any two intervals in a year, the peak load levels that occur in one are independent of those in the other. By ignoring the possibility of nonstationary, causal effects introduced by assigning units for maintenance outages, the same statistical independence may be asserted for the generating capacity states that occur in these two intervals. Thus

Expansion Analysis

the total availability and frequency of occurrence over the combined two-interval period of a specific margin state are, respectively, the sums of the availabilities and frequencies that occur in each of the separate intervals. This line of reasoning may be extended over all of the intervals making up the entire year. Equations 3.18 through 3.20 may be used to create annual values of the availability, frequency, and state transition rate if the kth state is taken as the annual value, the lth state is taken as denoting the value of the particular data describing this kth state in the lth interval, and N is the number of intervals considered to make up the year.

3.4 Expansion Analysis

3.4.1 Analysis of One Year

The examples given previously were for small or else idealized systems. A more realistic system of some 30 machines will be analyzed in detail to provide a comparison between the loss-of-load probability measure of generation reliability and the frequency and duration measures.

Table 3.14 lists the generation capacity data used for the example. Column four contains the repair-time data required to compute the frequency table. These average or mean repair times have all been selected as 2 days for the example. The other columns contain the data usually associated with a loss-of-load probability study.[20]

Table 3.14 30-Machine System

Number of Units	Unit Rating (MW)	Outage Rate (per unit)	Repair Rate (days)
9	10	0.0070	2.00
7	20	0.0070	2.00
2	40	0.0200	2.00
2	50	0.0200	2.00
4	90	0.0300	2.00
4	110	0.0300	2.00
2	220	0.0300	2.00
Total 30			

Application to Generation Planning   80

The annual load model data for the loss-of-load probability calculation are often taken as 260 weekday peak loads. Such a model will be used for the example. The expected load-peak duration for each day are the new data required for determining reserve margin frequency. (This is the value "e" used previously.) For this example, an average peak duration of 8 hours, or 0.333 day, was assumed for the year. The load model for the first example was adjusted to have an annual peak load of 1100 MW.

The probability, cycle time, and duration results for each maintenance period are shown in Table 3.15. In period 1, for example, a probability table has been computed for all the units not on maintenance, and it was used to determine the cumulative capacity outage probabilities associated with the 20 daily peak loads. For the loss-of-load probability index, these 20 probabilities are summed, resulting in 0.034004. This quantity may be interpreted as the expected or average number of days per period with available capacity below the load plus margin of 100 MW. Similarly for the zero margin condition, the expected number of days with a capacity shortage is 0.004268, and for the -100 MW margin, it is 0.000357. The zero margin state is used in determining when more capacity is needed. The three different margin states are used to determine the slope of the expectation characteristic and to indicate the sensitivity to load forecast uncertainties.

Using probability and frequency tables and the daily peak loads and durations for period 1 resulted in a cycle time of 153.7 periods and a duration of 0.219 day for the zero MW margin state. These quantities may be interpreted as the average or expected time between occurrences of events of available capacity less than the load and the average duration of the event, respectively.

The remainder of the columns for the interval summary present the data used in computing these three reliability measures. The available capacity was reduced to a 20-MW derating of the hydro units and the scheduled maintenance of 170 MW of capacity. The load was determined by forming the product of the 1100-MW annual peak forecast and the 0.8911 interval to annual ratio.

The program continues to study the 13 intervals in the year for a total of 260 days to produce the annual summary. The first column is the generation risk index referred to as "loss-of-load probability," which is

Expansion Analysis   81

equal to the sum of the 260 daily probabilities. The expected or average need for more capacity is indicated by the 0.067056 day/year associated with the zero margin state. The other two margins give an indication of the rate of change of the reliability measures and aid in assessing the sensitivity to load forecasting uncertainties. For example, an exponential curve is fitted through the coordinates 0.57 at 100 MW and 0.067 at zero MW and is used to estimate the annual peak load that would produce exactly 0.1 day/year, 1119 MW in this case.

With these data available, the frequency and duration results can be compared directly with the loss-of-load probability results. For example, at the zero margin state, the results indicate 0.067056 day/year (= annual LOLP) as the average number of days per year with a capacity shortage during the peak hour, 9.786 years as the cycle time between capacity shortages during the 8-hour average exposure period each day, and 0.219 day, or 5.25 hours, average duration of shortage.

These reliability measures are related in the following manner:

$$\begin{pmatrix} \text{Annual LOLP,} \\ \text{days/yr} \end{pmatrix} \begin{pmatrix} \text{Average Peak} \\ \text{Load Duration} \end{pmatrix} = \begin{pmatrix} \dfrac{\text{Average Shortage Duration, days}}{\text{Average Cycle Time, yr}} \end{pmatrix}.$$

For this specific example:

$$(0.067056)(0.333) = \begin{pmatrix} \dfrac{0.219}{9.786} \end{pmatrix}$$

$$0.0223 = 0.0223.$$

The two indices, frequency (or cycle time) and duration of shortage are the same as those used to describe components and systems in the transmission and distribution portions of a power complex. In addition, "availability" is frequently mentioned when discussing these components. Availability is the ratio of the mean uptime to the cycle time.[1] It is more easily computed by its relation to unavailability:

Table 3.15 Expansion Results

Interval Summary

| Period | Reserve Less Than Margin ||| Margin (MW) | Capability ||| Maint. Outage (MW) | Load (MW) |
	Average Number of Days/Period	Average Cycle Time (periods)	Average Duration (days)		Available (MW)	Derated (MW)			
1	0.034004	20.818	0.236	100.0	1460.0	−20.0	170.0	980.0
	0.004268	153.669	0.219	0.0				
	0.000357	1675.849	0.200	−100.0				
2	0.036272	19.498	0.236	100.0	1510.0	0.0	140.0	1043.0
	0.004655	140.823	0.219	0.0				
	0.000417	1437.233	0.200	−100.0				
3	0.035891	19.706	0.236	100.0	1520.0	0.0	130.0	1061.0
	0.004768	137.744	0.219	0.0				
	0.000402	1491.276	0.200	−100.0				
4	0.037494	18.788	0.235	100.0	1545.0	5.0	110.0	1092.0
	0.004824	135.594	0.218	0.0				
	0.000448	1335.340	0.200	−100.0				

...

Expansion Analysis

	Average Number of Days/Period	Average Cycle Time (periods)	Average Duration (days)	Margin (MW)	Total Capacity (MW)	Expected Load (MW)		
12	0.033922	20.865	0.236	100.0	1420.0	−20.0	170.0	964.0
	0.004437	148.230	0.219	0.0				
	0.000393	1530.034	0.200	−100.0				
13	0.034087	20.695	0.235	100.0	1460.0	−20.0	170.0	994.0
	0.004531	144.825	0.219	0.0				
	0.000394	1522.469	0.200	−100.0				
Total						−110.0		2160.0

Yearly Summary

Reserve Less Than Margin

Average Number of Days/Period	Average Cycle Time (periods)	Average Duration (days)	Margin (MW)	Total Capacity (MW)	Expected Load (MW)
0.573404	1.245	0.238	100.0	1650.0	1100.0
0.067056	9.786	0.219	0.0		
0.005257	113.702	0.199	−100.0		

System Capability at 0.10 Day/Yr

MW	Excess MW
1119.0	19.0

Application to Generation Planning

Availability (of Generation Surplus) = 1 − Unavailability

$$= 1 - \frac{\text{Duration, yr}}{\text{Cycle Time, yr}}$$

$$= 1 - \frac{0.219/365}{9.786} = 0.99994.$$

Note, in addition, that the probability at any time of a capacity shortage of as much as 100 MW may be estimated from Table 3.15, as

$$\frac{(\text{Average Shortage Duration, yr})}{(\text{Average Cycle Time, yr})} = \frac{(0.199/365)}{113.7} = 0.0000048.$$

3.4.2 Sensitivity

One of the interesting questions that may be investigated concerns the parameters for an interconnection. Suppose the 1650-MW system of Table 3.14 was to be interconnected to a much larger pool with a reserve of more than three times the rating of the tie. Consider a new tie rated at 220 MW for capacity planning purposes with an equivalent forced outage rate of either 0.0005 or 0.005 and with a mean repair time of 0.15 day. Table 3.16 summarizes the LOLP, frequency, and duration indices before and after the addition. No maintenance was considered for these computations to remove the effect of its variability from the study. The resulting "days/year" computed by the loss-of-load probability method are plotted for various load levels in Figure 3.6, while the cycle times in years are shown in Figure 3.7.

Assuming that the 1200-MW load level was acceptable for the 1650-MW generation system, then from the "day/year" graph in Figure 3.6, it can be seen that the 220-MW tie with a 0.0005 F.O.R. (Forced Outage Rate) increases the load-carrying capability of the system by 239 MW.* The capability increase is 224 MW with the 0.005 F.O.R. Both increases

---

* F.O.R. is the risk of component unavailability due to forced outage. For generating plants, F.O.R. is estimated by the ratio of forced outage time in hours to the sum of forced outage time plus service time. A term arising in electric utility planning, F.O.R. is an unavailability measure and should not be confused with measures of incidents per unit time.

Expansion Analysis

Table 3.16 Probability, Cycle Time, and Duration Results for Three Generating Systems

System I 1650-MW System

Annual peak load (MW)	1200.0
LOLP (day/yr)	0.0381
Cycle time (yr)	17.14
Duration (day)	0.217

System II 1650 MW + 220-MW Tie @ 0.0005

Annual peak load (MW)	1300.0	1500.0
LOLP (day/yr)	0.0022	0.1365
Cycle time (yr)	256.54	4.86
Duration (day)	0.185	0.221

System III 1650 MW + 220-MW Tie @ 0.005

Annual peak load (MW)	1300.0	1500.0
LOLP (day/yr)	0.0032	0.1623
Cycle time (yr)	127.92	3.38
Duration (days)	0.138	0.183

are larger than the rating of the tie because a load increase of 1 MW on the peak day represents only a percentage of that on every other day, while a capacity increase of 1 MW is 1 MW every day.

From the cycle time graph, Figure 3.7, it can be seen that the 220-MW tie with a 0.0005 F.O.R. and a repair time of 0.15 days produced a 238-MW increase in system capability, again assuming that the 1200-MW load level established the acceptability level for the system. The 220-MW tie with F.O.R. changed to 0.005 produced a 210-MW increase in system capability, a larger capability reduction on the cycle time graph, 28 MW, than was noted with the "days/year" measure, which indicated a 15-MW reduction.

The trip-out cycle times assumed for the tie in our example were

Application to Generation Planning

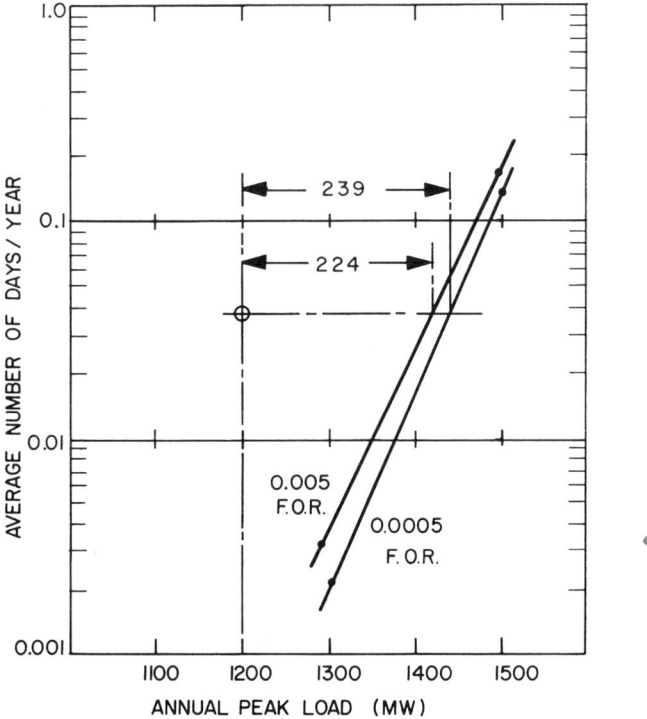

Figure 3.6 LOLP measure of tie effectiveness. 1650-MW system with 220-MW tie (0.15 day repair).

$$\text{Average Cycle Time} = \frac{\text{Repair Time}}{\text{Unavailability}}$$

$$= \frac{0.15}{0.005} = 30 \text{ days,}$$

for one case, and

$$\text{Average Cycle Time} = \frac{0.15}{0.0005} = 300 \text{ days,}$$

for the other case, and are for illustrative purposes only.

Expansion Analysis

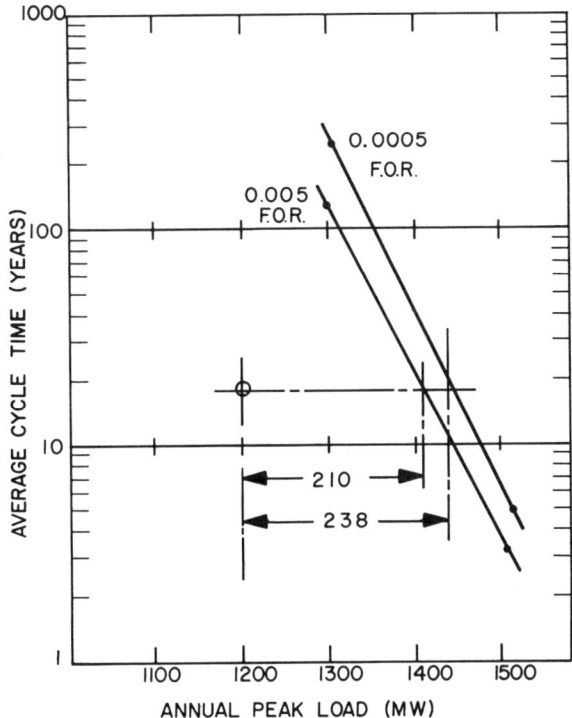

Figure 3.7 Cycle time measure of tie effectiveness. 1650-MW system with 220-MW tie (0.15 day repair).

The greater sensitivity of the cycle time index to transmission trip-out rates is of interest. This effect is not mere circumstance but rather is due to the recognition of component outage probability and outage duration parameters in the construction of the index. The LOLP index uses only the component outage probabilities. In this sense, the cycle time index may be superior to the LOLP index for interconnected system and bulk power-supply reliability evaluations. Note that the outage frequency and duration for transmission components differ markedly from the outage frequency and duration of generating units.

Of course, there are complicating factors in transmission-system analysis that are not accounted for in the preceding analysis. One is the dependency of the trip-out rate on the operating conditions. Other factors include such things as system behavior during large power

Table 3.17 Five-Year Expansion

Year	Run	Unit Changes (MW)	(F.O.R.)	Generating Capacity (MW)	Peak Load (MW)	Average Number of Days/Year	Average Cycle Time (yr)	Average Duration (days)
1	1	0	0.0	1650	1200	0.038063	17.1415	0.217
2	1	0	0.0	1650	1400	1.269989	0.5778	0.245
2	2	220	0.030	1870	1400	0.057383	11.3260	0.217
3	1	0	0.0	1870	1600	1.300611	0.5606	0.243
3	2	220	0.030	2090	1600	0.070620	9.1669	0.216
4	1	0	0.0	2090	1800	1.357424	0.5358	0.242
4	2	220	0.030	2310	1800	0.084220	7.6727	0.215
5	1	0	0.0	2310	2000	1.381889	0.5250	0.242
5	2	220	0.030	2530	2000	0.095092	6.7837	0.215

Expansion Analysis

transients and the handling of spinning reserve for the event of a tie trip.

3.4.3 Example Expansion

The system of Table 3.14 was expanded by adding 220-MW generating units. Each unit was assumed to have a forced outage rate of 0.03 and a mean repair time of 2 days. Again no maintenance was included so that the variability in the "days/year" and cycle time characteristics would result only from changes in load level and unit additions.

The summary of the resulting expansion is presented in Table 3.17. The 1650-MW capability system in Year 1 is the same as used for the previous tie investigation. With a new load level of 1400 MW, referred to as "Year 2," the reliability measures were computed, and these were used to plot the 1650-MW system characteristic in Figures 3.8 and 3.9. The computer was programmed to recognize that the resulting index, 1.27 days/year, exceeded the desired maximum of 0.1 and to automatically repeat the same load level with a 220-MW unit added. The resulting index of 0.057 day/year was satisfactory, and the study proceeded to the next load level.

With an annual peak load of 1600 MW, the program computed the second point on the (1650 + 220)-MW system characteristics in Figures 3.8 and 3.9. The effective load-carrying capability of the 220-MW unit, the distance in megawatts between the two "days/year" characteristics at the standard risk level, may now be measured in Figure 3.8. At the desired maximum of 0.1 day/year, the 1650-MW system could accept an annual peak load of 1254 MW, while the (1650 + 220)-MW system could accept 1437 MW, a difference of 183 MW. The 220-MW unit with a 3% F.O.R. is thus said to increase the system's load-carrying capability by 183 MW.

The procedure of studying a load level, adding a 220-MW unit and restudying, and then moving to a level 200 MW higher was repeated until the level of 2000 MW was reached with four 220-MW units added. The effective capabilities of these units are shown in Figure 3.8, increasing from 183 MW to 193 MW. This progression in capability with the number of identical units is to be expected.[21]

What cycle time corresponds with the 0.1 day/year desired value? For the 1650-MW system, 1254 MW of load would give 0.1 day/year, and from Figure 3.9, 1254 MW of load would give 6.8 years between

Application to Generation Planning 90

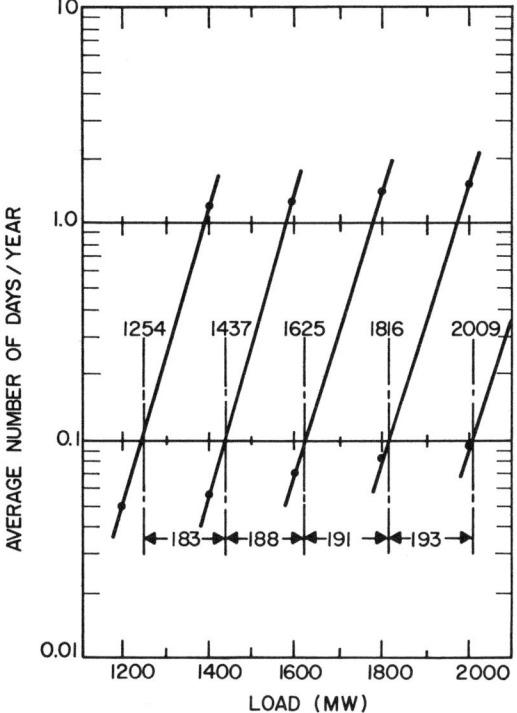

Figure 3.8 LOLP measure of generation. Effective capability of 220-MW units with 3% F.O.R. and 2 days mean repair time.

occurrences. Measuring the effect of unit additions at this 6.8-years level produces the effective capabilities of 180 to 192 MW. These are encouragingly close to those derived from the loss-of-load probability method. Because these effective capabilities are slightly smaller, they do indicate that the cycle time is shrinking slightly as the "days/year" is held constant. The cycle-time measure appears to be a bit more conservative an estimate of a system's change in reliability with the increase in capacity.

3.5 Load Statistics

The demand model assumes that the duration of the daily peak load is an exponentially distributed random variable. Several sets of actual

Load Statistics

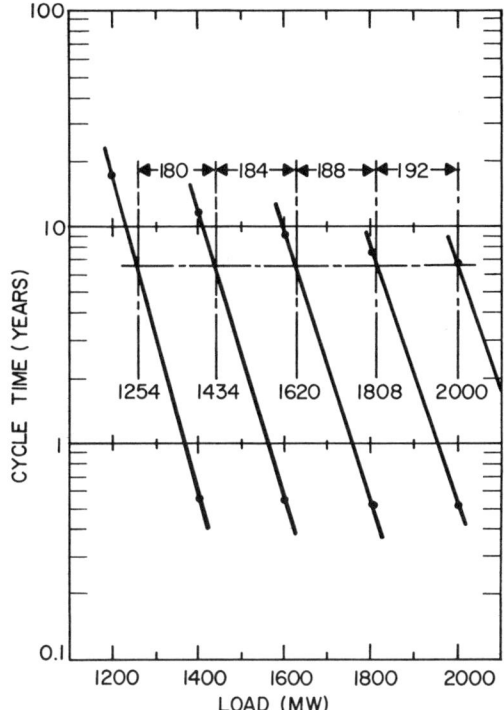

Figure 3.9 Cycle time measure of generation. Effective capability determined by cycle time for 220-MW units with 3% F.O.R. and 2 days mean repair time.

hourly load data have been examined to consider this assumption. The results are now discussed briefly.

This load model is used for evaluating generation installed reserve requirements. Therefore, it appears logical that the mean duration of this peak load should span the entire peak load period, which is approximately 8 hours in Figure 3.10. This would make the load exposure factor e have a value of $\frac{1}{3}$.

The question of whether or not the load duration is in fact distributed exponentially is of interest but of relatively minor importance as long as the load duration distribution has a finite mean value. This is because the reliability analysis deals only with mean values, and the results, which are also mean values, are independent of the distribution. The exponential distribution does, however, fit actual load duration dis-

Application to Generation Planning

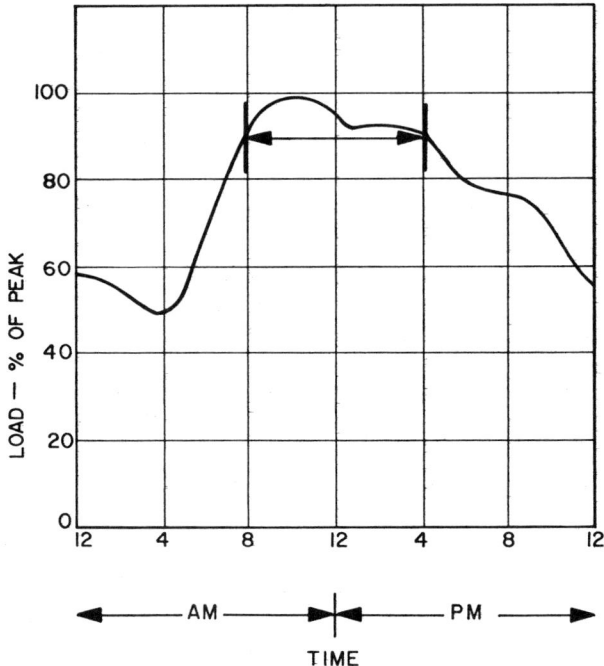

Figure 3.10 Definition of peak load hours. Duration = 8 hours at 90% load level.

Table 3.18 Characteristics of Four Power Companies

Company	Length of Data (yr)	Characteristics
"B"	5	Northeastern metropolitan area.
"L"	5	Northeast location. Primarily suburban with some industry.
"M"	5	Southern location. Large area with diverse load types.
"T"	3	Southwestern metropolitan utility.

Load Statistics 93

tributions for limited ranges of duration for the actual load data examined.

3.5.1 Analysis
The load histories of four different types of U.S. electric utility companies were examined and analyzed. For each company, 3 or 5 years of hourly load records were available. The data are arranged chronologically, and each day's record is identified as to date and day of the week. The four utilities and some of their attributes are given in Table 3.18.

The load data were examined to find the distribution of the duration of "peak load." Peak loads are classified in percentage of the daily peak. Figure 3.10 indicates the method of classification. For example, the load level of 90% of the daily peak had a duration of 8 hours. All hours were examined, and the analysis resulted in

1. distributions of daily peak durations for a given load level, and
2. mean durations as a function of the load level.

The daily peak-load duration distribution is the probability that a given load level will exist for a given time or longer. This statistic must go through 1.00 at a duration of 1 hour since the smallest recorded time interval is 1 hour. The load-level classification in percentage of daily peak is appropriate for this load model since the distribution of the magnitude of daily peak occurrences may be forecast separately.

Each company's data were classified by:

1. day types (i.e., weekdays, weekends and holidays, and all days), and

2. seasons:
   winter: December, January, February;
   spring: March, April, May;
   summer: June, July, August;
   fall: September, October, November.

The load levels considered ranged from 98% of the daily peak to 76%, and data were accumulated in 2% steps.

3.5.2 Results
Figure 3.11 shows the mean durations of the weekdays for all four companies and all seasons of the year for various peak load levels. For a mean duration of 8 hours, the load levels for the four companies are 83% to 82% of the daily peak. The data for company "M" indicates a

Application to Generation Planning 94

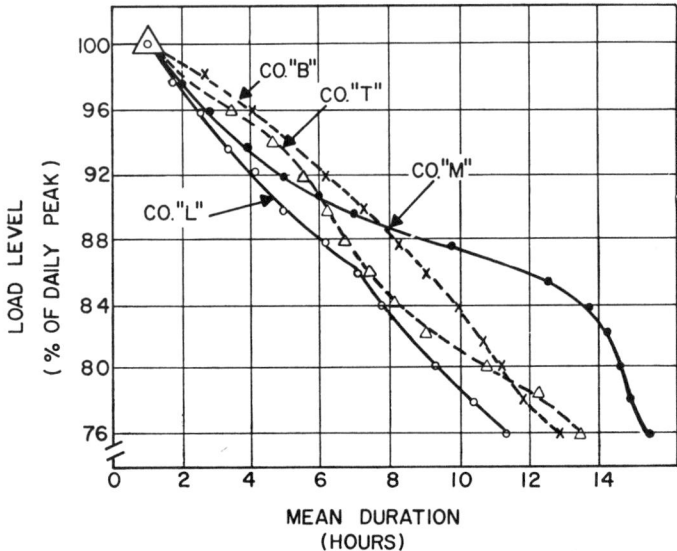

Figure 3.11 Weekday load levels versus duration.

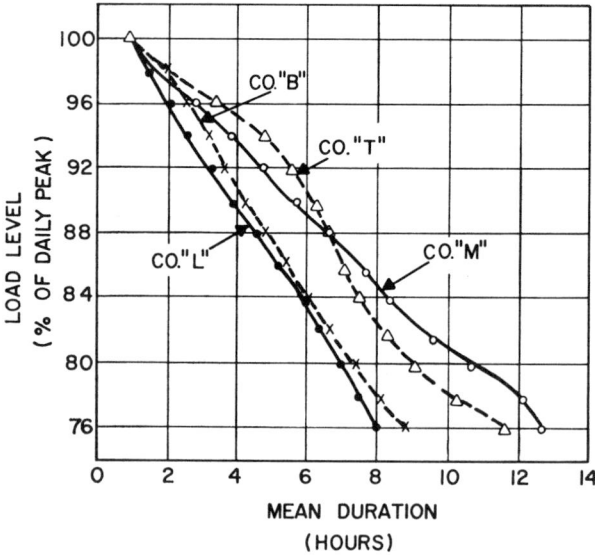

Figure 3.12 Nonweekday load levels versus duration.

Load Statistics 95

weather-sensitive, load-duration characteristic, the large bulge being due primarily to the summer. Figure 3.12 gives the same characteristics for the weekends and holidays.

Figure 3.13 plots the company "M" weekday and weekend peak load duration distribution for peak load levels for which the mean duration is closest to 8 hours. Also shown is the exponential distribution with a mean duration of 8 hours. These statistical data indicate that the exponential distribution is a reasonably good approximation. A further indication of this is shown in Figure 3.14. This is a plot of the data points and approximately fit exponential variations for a number of load levels and for all of the (unclassified) data for company "M." The straight lines representing exponential distributions were drawn through a probability of 1.0 at a load duration of 1 hour since all of the actual data curves must go through this point.

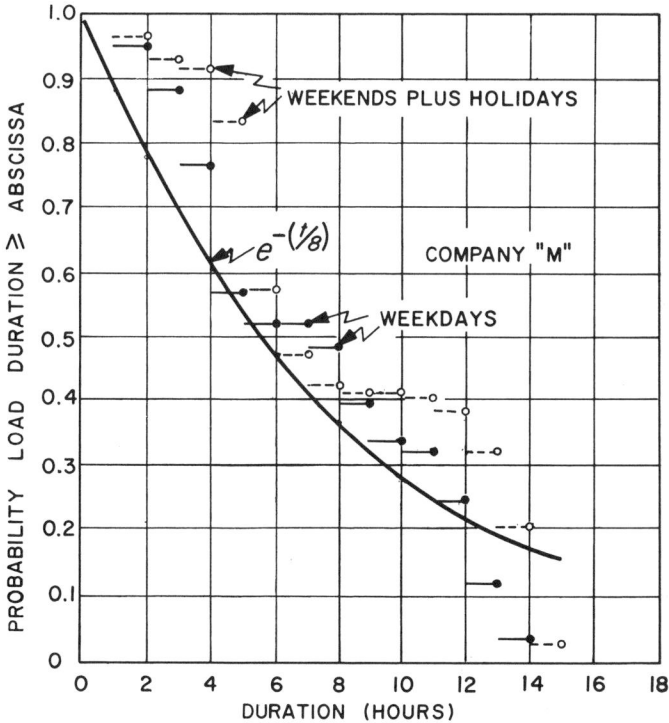

Figure 3.13 Distribution of load durations.

Application to Generation Planning 96

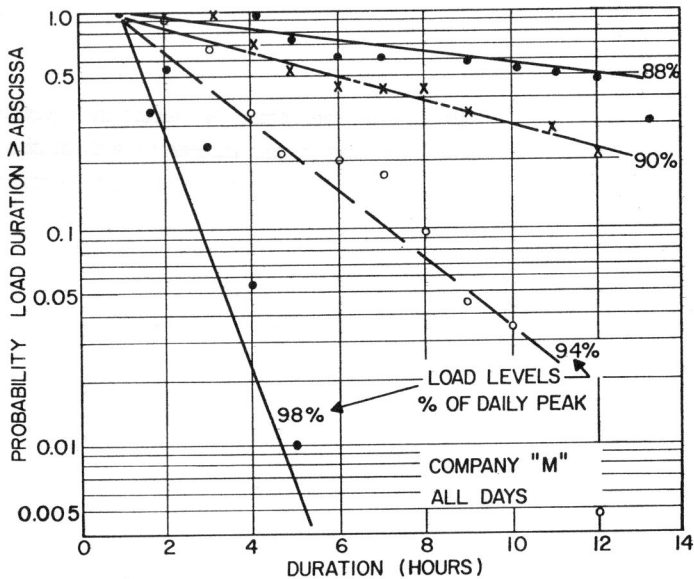

Figure 3.14 Semilog graph of distributions of load durations.

3.5.3 Mean Duration Estimates

From a brief examination of 18 years of hourly load history for four different types of utilities, it is concluded that an exponential distribution for the duration of daily peak loads is a good approximation to the actual distributions. For planning generation reserves, it would appear that a mean peak-load duration of 8 hours would be appropriate. However, in areas such as the southern United States, where the weather may cause peak-load conditions to persist longer, a mean duration of 10 to 12 hours may be more appropriate. Thus, the load exposure factor e would be between 0.333 and 0.500.

3.6 Interpretation of the LOLP and Margin Frequency Indices

An algebraic comparison of the LOLP and the margin frequency indices is of assistance in comparing and contrasting them. The algebraic expression for the LOLP index may be written as

Interpretation of the LOLP and Margin Frequency Indices

$$\text{LOLP} = \sum_L n_L A_G, \quad (3.32)$$

where n_L is the expected number of occurrences of load L during the year and A_G is the corresponding unavailability of generation capacity sufficient to serve load L.

In a similar way, one may obtain the expressions for the margin frequency and probability:

$$f_m = \sum_L A_L [f_G + A_G(\lambda_{-L} - \lambda_{+L})] \quad (3.33)$$

and

$$A_m = \sum_L A_L A_G, \quad (3.34)$$

where A_L is the availability of load L, $\lambda_{+L}(\lambda_{-L})$ is the load transition rate to higher (lower) load magnitudes, and f_G is the frequency with which generation capability insufficient to serve load L will occur.

Expressions relating A_L and n_L are offered in the Table 3.19. These apply for a year of 365 days.

The probability (availability) of zero margin A_m may then be written as

$$A_m = \frac{e}{365} \sum_L n_L A_G + (1 - e) A_{G_0}, \quad (3.35)$$

Table 3.19 Expressions Relating A_L and n_L

Daily Load Condition	Annual Occurrences	Duration	A_L	λ_{+L}	λ_{-L}
Peak—Load L	n_L	e	$e\left(\dfrac{n_L}{365}\right)$	0	$\dfrac{1}{e}$
Low Load	365	$1 - e$	$1 - e$	$\dfrac{1}{1 - e}$	0

Application to Generation Planning

where A_{G_0} is the probability of a generation deficiency during the low-load period. The frequency of the zero margin condition may be written as

$$f_m = \sum_L A_L f_G - A_{G_0} + \frac{1}{365} \sum_L n_L A_G + (1 - e)f_{G_0} \text{ per day}, \quad (3.36)$$

but by the definition of the LOLP index

$$f_m = \sum_L A_L f_G - A_{G_0} + \left(\frac{1}{365}\right)(\text{LOLP}) + (1 - e)f_{G_0} \text{ per day}, \quad (3.37)$$

where f_{G_0} is the frequency of a generation deficiency occurring during a low-load period.

In most practical cases, the risk during the low-load period is very small compared to the peak-load period:

$$f_{G_0} \ll A_L f_G$$

and

$$A_{G_0} \ll A_G;$$

hence,

$$f_m \approx \sum_L A_L f_G + \frac{1}{365}(\text{LOLP})$$

$$= e \sum_L n_L f_G + \frac{1}{365}(\text{LOLP}) \text{ per day.} \quad (3.38)$$

For very small e, note that LOLP/365 and f_m are the same and that the margin frequency index and the LOLP index become equivalent if the margin frequency is converted to an annual basis.

References

1. J. D. Hall, R. J. Ringlee, and A. J. Wood, "Frequency and Duration Methods for Power System Reliability Calculations, Part I—Generation System Model," *IEEE, Transactions on Power Apparatus and Systems*, vol. 87, 1968, pp. 1787–1796.

2. R. J. Ringlee and A. J. Wood, "Frequency and Duration Methods for Power System Reliability Calculations, Part II—Demand Model and Capacity Reserve Model," *IEEE, Transactions on Power Apparatus and Systems*, vol. 88, 1969, pp. 378–388.

3. C. D. Galloway, L. L. Gaver, R. J. Ringlee, and A. J. Wood, "Frequency and Duration Methods for Power System Reliability Calculations, Part III—Generation System Planning," *IEEE, Transactions on Power Apparatus and Systems*, vol. 88, 1969, pp. 1216–1223.

4. V. M. Cook, R. J. Ringlee, and A. J. Wood, "Frequency and Duration Methods for Power System Reliability Calculations, Part IV—Models for Multiple Boiler-Turbines and for Partial Outage States," *IEEE, Transactions on Power Apparatus and Systems*, vol. 88, 1969, pp. 1224–1232.

5. R. J. Ringlee and A. J. Wood, "Frequency and Duration Methods for Power System Reliability Calculations, Part V—Models for Delays in Unit Installations and Two Interconnected Systems," *IEEE, Transactions on Power Apparatus and Systems*, vol. 90, 1971, pp. 79–88.

6. H. Halperin and H. A. Adler, "Determination of Reserve Generating Capability," *AIEE Transactions*, vol. 77, pt. III, 1958, pp. 530–544.

7. AIEE Subcommittee Report, "Application of Probability Methods to Generating Capacity Problems," *AIEE Transactions*, vol. 79, pt. III, 1960, pp. 1165–1182.

8. V. M. Cook, C. D. Galloway, M. J. Steinberg, and A. J. Wood, "Determination of Reserve Requirements of Two Interconnected Systems," *IEEE, Transactions on Power Apparatus and Systems*, vol. 82, 1963, pp. 18–33.

9. C. F. DeSieno and L. L. Stine, "A Probability Method for Determining the Reliability of Electric Power Systems," *IEEE, Transactions on Power Apparatus and Systems*, vol. 83, 1964, pp. 174–181.

10. D. P. Gaver, F. E. Montmeat, and A. D. Patton, "Power System Reliability: I—Measures of Reliability and Methods of Calculation," *IEEE, Transactions on Power Apparatus and Systems*, vol. 83, 1964, pp. 727–737.

11. F. E. Montmeat, A. D. Patton, J. Zemkoski, and D. J. Cummings, "Power System Reliability: II—Applications and a Computer Program," *IEEE, Transactions on Power Apparatus and Systems*, vol. 84, 1965, pp. 636–643.

12. S. A. Mallard and V. C. Thomas, "A Method for Calculating Transmission System Reliability," *IEEE, Transactions on Power Apparatus and Systems*, vol. 87, 1968, pp. 824–834.

13. R. J. Ringlee, "Steps in Measuring Service Continuity," *Distribution* (General Electric Co., Schenectady, N.Y.), vol. 28, Fourth Quarter, 1966, pp. 14–18.

14. M. A. Sager, R. J. Ringlee, and A. J. Wood, "A New Production Cost Program to Recognize Forced Outages," IEEE Paper No. TP 72 159-7, 1972 Winter Power Meeting.

15. M. A. Sager and A. J. Wood, "Power System Production Cost Calculations Recognizing Forced Outages," IEEE Paper No. TP 72 158-9, 1972 Winter Power Meeting.

16. S. J. Einhorn, "Reliability Prediction for Repairable Redundant Systems," *Proceedings of the IEEE*, vol. 51, 1963, pp. 312–317.

17. D. M. Sauter, C. J. Baldwin, and K. M. Dale, discussion of Halperin and Adler (6), pp. 541–542.

18. E. L. Arnoff and J. C. Chambers, "Operations Research Determination of Generator Reserves," *AIEE Transactions*, vol. 76, pt. III, 1957, pp. 316–328.

19. Edison Electric Institute, "Report on Equipment Availability for the Seven Year Period 1960–66," EEI Publication 67-23.

20. C. D. Galloway and L. L. Garver, "Computer Design of Single-Area Generation Expansions," *IEEE, Transactions on Power Apparatus and Systems*, vol. 83, 1964, pp. 305–311.

21. L. L. Garver, "Effective Load-Carrying Capability of Generating Units," *IEEE, Transactions on Power Apparatus and Systems*, vol. 85, 1966, pp. 910–919.

The following references are relevant although not cited in the chapter.

22. A. Papoulis, *Probability, Random Variables and Stochastic Processes*. New York: McGraw-Hill, 1965.

23. R. Billinton and K. E. Bollinger, "Transmission System Reliability Evaluating Using Markov Processes," *IEEE, Transactions on Power Apparatus and Systems*, vol. 87, 1968, pp. 538–547.

24. R. Billinton, "Composite System Reliability Evaluation," *IEEE, Transactions on Power Apparatus and Systems*, vol. 88, 1969, pp. 276–280.

25. J. G. Kemeny and J. L. Snell, *Finite Markov Chains*. Princeton, N.J.: Van Nostrand, 1959.

4 APPLICATIONS TO BULK POWER-SUPPLY SYSTEMS

4.1 Introduction

The previous chapter illustrates the determination of generating capacity reliability indices in single systems. The concepts outlined can be applied directly to more than one system if there are no transmission limitations between the interconnected areas. The approach must be modified, however, if there are transmission restrictions upon the possible available assistance. This situation may also exist within a single system that contains two finite generating capacity areas connected by transmission facilities. The total system reliability and the individual area reliabilities are therefore influenced by any restrictions due to the interconnecting transmission.

As previously noted in Chapter 1, Cook and his co-workers[1] introduced the basic idea of an array containing the probabilities of various capacity conditions within the two regions. The transmission capability between the two areas can then be superimposed upon the array and the individual-system capacity-outage probabilities modified accordingly. This technique can be extended to an approach in which the interconnected system is seen as an equivalent capacity assistance probability table.[2] The reliability indices obtained using this approach are the conventional loss-of-load expectation parameters. The recursive approach to obtaining frequency and duration indices of capacity adequacy has also been extended to the multiarea problem.[3-5] Indices of availability, frequency, and duration of a negative margin condition in which the load exceeds the available capacity by a given amount can be determined for each area and include the possible assistance from neighboring areas.

These three indices can also be determined for any load point within a system by a composite reliability approach. This technique utilizes a quality of service rather than a continuity of service criterion. A system failure is charged if the supply at a major transmission bus does not meet predetermined voltage standards. Equipment overloads and cascading outages due to thermal limitations can be included in this approach.

Applications to Bulk Power-Supply Systems 102

The two-area reliability approach and the composite system technique are illustrated by applications in this chapter.

4.2 Two-Area Reliability Evaluation

The basic concepts for capacity reliability evaluation illustrated in Chapter 3 can be easily extended to the two-area problem.[4] By using a similar approach to that illustrated previously, a two-dimensional margin state array can be created, as shown in Figure 4.1 for two hypothetical systems, A and B. In effect, there are two separate sets, one that exists when the interconnection is available and one that exists when the interconnection is out of service. If the interconnection is assumed to be completely reliable, then only the left-hand array exists. As in the single-system case, the capacity and the load in each system are assumed to exist at a discrete number of levels and therefore the margin states also exist at discrete levels.

Considering System A, the margin state m_{ij} in the left-hand array is given by

$$m_{ij} = m_{ai} + h_{ij},$$

where

h_{ij} = the assistance available to A from B

and

m_{ai} = the individual margin in A.

If the interconnection is out of service then

$$m_{ij} = m_{ai}.$$

Recursive relationships are available for computing the availability and frequency of the margin states. As indicated on page 74 in the single-system case, it is highly desirable to compute the cumulative indices directly, and algorithms are available for this purpose.

Two-Area Reliability Evaluation   103

INTERCONNECTION UP

	M_b							
	m_{b1}	m_{b2}	m_{b3}	m_{b4}	•	m_{bNB}		
M_a								
m_{a1}	m_{11}	m_{12}	m_{13}	m_{14}	•	m_{1NB}		
m_{a2}	m_{21}	m_{22}	m_{23}	m_{24}	•	m_{2NB}		i
m_{a3}	m_{31}	m_{32}	m_{33}	$^h m_{34}$	•	m_{3NB}		g
m_{a4}	m_{41}	m_{42}	$^f m_{43}$	m_{44}	•	m_{4NB}		d
m_{a5}	m_{51}	$^c m_{52}$	m_{53}	m_{54}	•	m_{5NB}		j
m_{a6}	m_{61}	m_{62}	m_{63}	m_{64}	•	m_{6NB}		
	•	•	•	•		•		
b	•	•	•	•		•		
m_{aNA}	m_{NA1}	m_{NA2}	m_{NA3}	m_{NA4}	•	m_{NANB}		e
a								

INTERCONNECTION DOWN

	M_b						
	m_{b1}	m_{b2}	m_{b3}	m_{b4}	•	m_{bNB}	
M_a							
m_{a1}	m'_{11}	m'_{12}	m'_{13}	m'_{14}	•	m'_{1NB}	k2
m_{a2}	m'_{21}	m'_{22}	m'_{23}	m'_{24}	•	m'_{2NB}	k1
m_{a3}	m'_{31}	m'_{32}	m'_{33}	m'_{34}	•	m'_{3NB}	k3
m_{a4}	m'_{41}	m'_{42}	m'_{43}	m'_{44}	•	m'_{4NB}	
m_{a5}	m'_{51}	m'_{52}	m'_{53}	m'_{54}	•	m'_{5NB}	
m_{a6}	m'_{61}	m'_{62}	m'_{63}	m'_{64}	•	m'_{6NB}	
	•	•	•	•		•	
	•	•	•	•		•	
m_{aNA}	m'_{NA1}	m'_{NA2}	m'_{NA3}	m'_{NA4}	•	m'_{NANB}	

(labels: j2, j1, j3 near m_{a1}, m_{a2}, m_{a3} in the down matrix)

Figure 4.1 Effective margin states matrices for System A connected to System B. The thick lines denote ± margin states.

Table 4.1 Load Model for Systems A and B

Exposure factor = 0.5 day
Period = 20 days

Load Level (MW)	No. of Occurrences
1450	8
1255	4
1155	4
1080	4

The low-load level was assumed to be zero.

This approach has been applied to the 22-unit generating system described on page 61. It is assumed that both Systems A and B are identical and have this composition. The total installed capacity of each system is therefore 1725 MW. Each system was assumed to have the load model indicated in Table 4.1.

The availability and cycle time indices of capacity adequacy[5] in System A as a function of the tie-line transfer capability are shown in Figures 4.2 and 4.3. The tie-line failure and repair rates were taken as 0.01 failure/day and 2.5 repairs/day, respectively. As noted previously, the availability index in Figure 4.2 could be directly converted to a loss-of-load probability, or, more correctly, expectancy. The limiting tie capacity is defined as the tie capacity beyond which there results no further increase in reliability benefit. As can be seen from Figures 4.2 and 4.3, this occurs at approximately 225 to 250 MW. From a practical point of view the incremental reliability benefits are very small for tie capacities in excess of 300 MW. The shapes of these two curves are basically identical. The loss-of-load and the frequency-and-duration approaches, however, react in different ways to variation in the tie-line forced outage rate, as shown in Figures 4.4 and 4.5. As in the case of single-system studies, the frequency-and-duration approach does provide a more physical interpretation of the system reliability and is more amenable to sensitivity studies of the component failure parameters.

4.3 Composite System Reliability Evaluation

Power-system transmission requirements and their associated costs are influenced considerably by the location of the existing and proposed

Composite System Reliability Evaluation

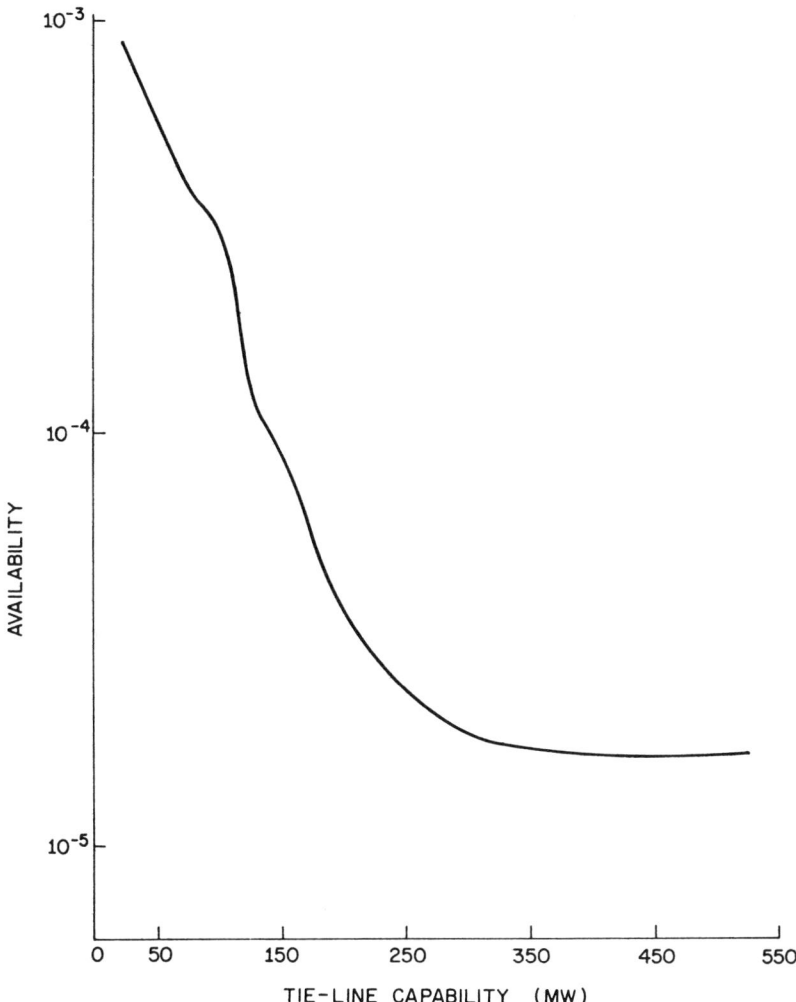

Figure 4.2 Variation of risk level (availability) in System A with the variation of tie-line capability.
Peak in System A = 1450 MW
Peak in System B = 1450 MW
Failure rate of tie line = 0.01 failure/day
Repair rate of tie line = 2.5 repairs/day

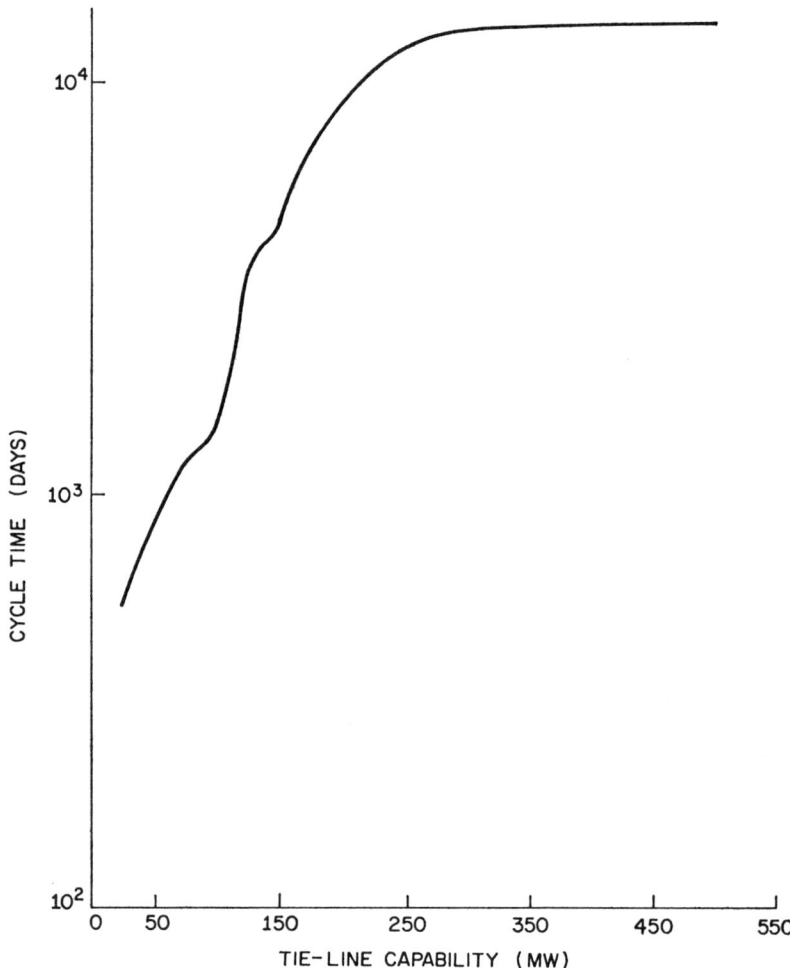

Figure 4.3 Variation of risk level (cycle time) in System A with the variation of tie-line capability.
Peak in System A = 1450 MW
Peak in System B = 1450 MW
Failure rate of tie line = 0.01 failure/day
Repair rate of tie line = 2.5 repairs/day

Composite System Reliability Evaluation

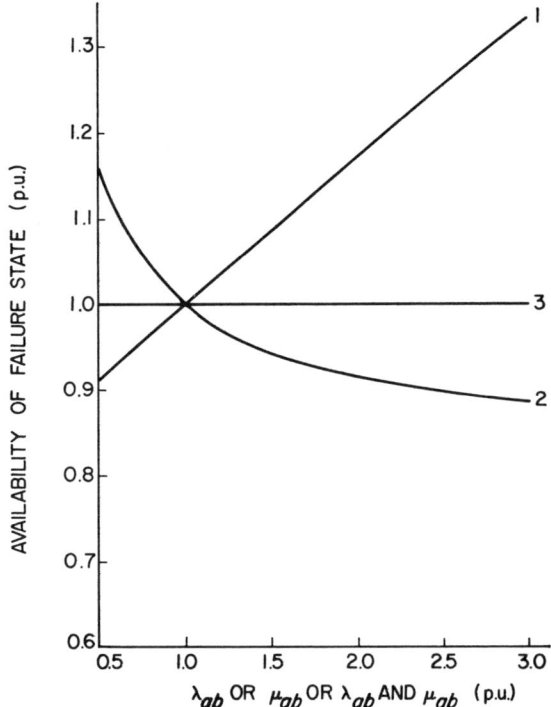

Figure 4.4 Variation in risk level (availability) of System A with variations in tie-line mean failure and mean repair rates.
Tie-line capability = 200 MW
Base λ_{ab} = 0.01 failure/day
Base μ_{ab} = 2.5 repairs/day
Peak in System A = 1450 MW
Peak in System B = 1450 MW
1. Only λ_{ab} varied keeping μ_{ab} at 1.0 p.u.
2. Only μ_{ab} varied keeping λ_{ab} at 1.0 p.u.
3. Both λ_{ab} and μ_{ab} varied in the same ratio

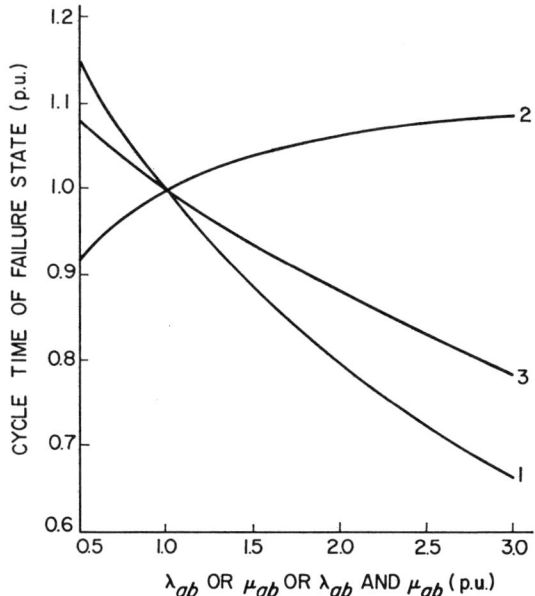

Figure 4.5 Variation in risk level (cycle time) of System A with variation in tie-line mean failure and mean repair rates.
Tie-line capability = 200 MW
Base λ_{ab} = 0.01 failure/day
Base μ_{ab} = 2.5 repairs/day
Peak in System A = 1450 MW
Peak in System B = 1450 MW
1. Only λ_{ab} varied keeping μ_{ab} at 1.0 p.u.
2. Only μ_{ab} varied keeping λ_{ab} at 1.0 p.u.
3. Both λ_{ab} and μ_{ab} varied in the same ratio

generating capacity. The selection of future capacity installations may require the examination of many alternatives. These alternatives are amenable to economic comparison provided that they all meet some design level of reliability for the system in question. The application of probability methods to generating-capacity reliability evaluation was illustrated in the previous chapter, in which the total system load and the system generation were considered as individual elements. The assumption was made that adequate transmission exists in the system to move the generated power to the load points. The reliability at a partic-

Conditional Probability Approach

ular load point is dependent upon the total installed capacity but is also directly related to the position of the load point within the system and the available connecting transmission facilities. A method has been described of evaluating the reliability at any point in a composite system including both generating and transmission facilities.[6] This approach utilizes a service quality standard as the reliability criterion rather than simple continuity between sources and load points. A general design criterion was postulated in which "if with the various possible combinations of system components out of service, the reliability index for the station is below an acceptable minimum, then additional facilities are required to meet the reliability standards." The effect of failures during storms, acceptable bus-voltage levels, shunt capacitive compensation, and transmission redundancy have been illustrated for a simple system. The basic technique for analyzing a more complicated network was also illustrated. This approach can be readily incorporated in a digital computer program for system transmission development and alternate expansion plan studies. The approach breaks down into two separate phases, the evaluation of the reliability index in a practical system network and the logical addition of subsequent facilities.

4.4 Conditional Probability Approach

The reliability analysis can be performed using a very simple but powerful probability concept.

If two events A and B are independent,

$$P(A \cap B) = P(A) \cdot P(B).$$

If the events are not independent and $P(B \mid A)$ is the conditional probability that B occurs, given that A has occurred,

$$P(A \cap B) = P(A) \cdot P(B \mid A).$$

If the occurrence of A is dependent upon a number (j) of events B_i that are mutually exclusive,

$$P(A) = \sum_{i=1}^{j} P(A \mid B_i) P(B_i). \tag{4.1}$$

Applications to Bulk Power-Supply Systems

If the occurrence of A is dependent upon only two mutually exclusive events for component B, success and failure, designated B_x and B_y, respectively,

$$P(A) = P(A \mid B_x) \cdot P(B_x) + P(A \mid B_y) \cdot P(B_y).$$

With regard to reliability this can be expressed in a more direct form:

$P(\text{system failure}) = P(\text{system failure if } B \text{ is good}) \cdot P(B_x)$
$\qquad\qquad\qquad\qquad + P(\text{system failure if } B \text{ is bad}) \cdot P(B_y).$

In a power-system network there are a number of possible outage combinations of lines, transformers, and generating units. Each outage condition has a probability of existence and a frequency of occurrence. Under each outage condition there is a maximum load at each bus that can be supplied without violating the service quality criterion. The probability that the load will exceed this maximum can be determined from the load probability distribution for the bus in question. This is a conditional probability, since the maximum load is determined given that a certain outage condition has occurred. The system generating facilities can be included by developing a capacity outage probability table for all the units within the system.[6] The maximum load that can be supplied at a bus can be obtained for any given outage condition in the transmission system. The probability of the load at the bus exceeding this maximum value is then combined with the probability that the available system generating capacity is insufficient to meet the total system load. Generation and transmission outage conditions are considered as two independent events resulting in failure at a bus.

Consider the case of a generating station supplying a load over two identical transmission lines.

P_G = probability that the load will exceed the available generating capacity.

P_L = probability that the bus load will exceed the available transmission capability.

That is:

$P_{L_1} = P(\text{load exceeding the 2 line capability})$,
$P_{L_2} = P(\text{load exceeding the 1 line capability})$,
$P_{L_3} = P(\text{load exceeding the 0 line capability}) = 1.0$.

Conditional Probability Approach

A_L = line availability,
$\bar{A}_L$ = line unavailability.

Given that both lines are in:

$$Q_s = P(\text{load point failure}) = P_G + P_{L_1} - P_G \cdot P_{L_1}.$$

Given that one line is in:

$$Q_s = P_G + P_{L_2} - P_G \cdot P_{L_2}.$$

Given that both lines are out:

$$Q_s = P_G + P_{L_3} - P_G \cdot P_{L_3} = 1.0.$$

The probability that both lines are in (i.e., State 1) = A_L^2.

$P(\text{State 2}) = P(\text{one line in}) = 2A_L\bar{A}_L.$

$P(\text{State 3}) = P(\text{both lines out}) = \bar{A}_L^2.$

∴ Using the conditional approach but with three mutually exclusive events:

$$Q_s = A_L^2(P_G + P_{L_1} - P_G \cdot P_{L_1}) + 2A_L\bar{A}_L(P_G + P_{L_2} - P_G \cdot P_{L_2}) + \bar{A}_L^2.$$

The equation for load-point unavailability can be expressed in a general form:

Probability of failure at bus k

$$= \sum_j P(B_j)(P_{G_j} + P_{L_j} - P_{G_j} \cdot P_{L_j}), \quad (4.2)$$

where
  B_j = a state in the transmission network (line and transformers),
  $P(B_j)$ = probability of existence of state B_j,
  P_{G_j} = probability of the generating capacity outage exceeding the reserve capacity (a cumulative probability figure obtained from the capacity outage probability table), and

Applications to Bulk Power-Supply Systems

P_{L_j} = probability of load at bus k exceeding the maximum load that can be supplied at that bus without failure.

The probability of existence of state B_j is obtained by assuming that the individual component outages are independent. The effect of storm-associated failures on the probability of system failure is virtually negligible in those cases in which there is a reasonable probability that the system load will exceed the remaining transmission capability.[2] The effect of storms on reliability evaluation of major transmission over relatively long distances would be further diminished as storms become more local in nature.

It is also possible to determine a reliability index at any bus in the system in terms of an average or expected frequency of failure. The frequency of occurrence of an outage condition is equal to the product of the probability of existence of the outage and the rate of departure from that condition.[2] If the generating unit outages and the load variation are considered in terms of probability only and not in terms of frequency of occurrence, then the expected frequency of failure at bus k is given by

$$\sum_j [F(B_j)(P_{G_j} - P_{G_j} \cdot P_{L_j})], \qquad (4.3)$$

where

$F(B_j)$ = frequency of occurrence of outage B_j.

If the rates of departure associated with the individual generating-unit states and the load-model states for each bus in the system are included, the evaluation becomes extremely complicated.

In the approach described by Equation 4.2, the generation schedule used in the load-flow analysis is not modified to include the outage of individual units. The assumption is made that adequate generation is available and any breach of quality is due to line or transformer outages or to a shortage of generation capacity available to meet the system load. This assumption is not required if the generating units are considered individually, together with the transmission lines and transformers, to determine each outage condition B_j. The generation schedule is then modified for each generation outage condition. The number of individ-

Conditional Probability Approach

ual outage conditions in this case may be much greater than those considered using Equation 4.2. This approach is, however, more accurate since the bus voltage and line loadings are affected by the generation schedule.

The equations for this case are as follows:

$$\text{Probability of failure at bus } k = \sum_j P(B_j) \cdot P_{L_j}. \tag{4.4}$$

$$\text{Expected frequency of failure at bus } k = \sum_j F(B_j) \cdot P_{L_j}. \tag{4.5}$$

In Equation 4.5 as in Equation 4.3, the rates of departure associated with the load-model states are not included in the evaluation.

The load model used in Equations 4.4 and 4.5 is the conventional cumulative distribution of daily peak loads or hourly loads for the period under study. As illustrated in a paper by Billinton and Bhavaraju,[7] it is possible to utilize a load model of the form illustrated in Figure 3.5 and Table 3.11. For each load level L, there are certain outage states for which the system will be able to supply only a part of the load due to limited line capabilities. These states can be termed as negative margin states for the specific load level. The usual loss-of-load probability approach in generation planning considers all negative margins as loss of load and does not distinguish between the different negative margin magnitudes. A simplified expression to obtain the cumulative frequency of negative margin states utilizing the probability and frequency of the various states was given earlier as Equation 3.33. The same basic expression has also been used in transmission-system reliability evaluation by interpreting the generating-capacity outage states of Chapter 3 as outage states j in Equations 4.4 and 4.5.

$$\text{Expected loss of load at a bus} = \sum_L n_L \cdot A_G \text{ days/period.} \tag{4.6}$$

(Compare with Equation 3.32.)

$$\text{Frequency of loss of load} = (1 - e)[f_{G_0} + A_{G_0}(\lambda_{-L_0} - \lambda_{+L_0})]$$
$$+ \sum_L A_L[f_G + A_G(\lambda_{-L} - \lambda_{+L})], \tag{4.7}$$

Applications to Bulk Power-Supply Systems   114

where

f_G = the cumulative frequency of the outage states that result in a negative margin for load level L, corrected for the frequency of transfers within the states,

f_{G_0} = a similar value for the low-load condition,

A_G = the cumulative probability of the outage states that result in a negative margin for load level L,

and

A_{G_0} = a similar value for the low load.

If the risk in the low-load period can be neglected, this portion of Equation 4.7 can be neglected and the expression reduces to

$$\text{frequency of loss of load} = \sum_L A_L [f_G + A_G(\lambda_{-L} - \lambda_{+L})].$$

Substituting for the load transfer rates and multiplying by the number of days in the period, the frequency of load loss at a bus in occurrences per period is given by

$$D \sum_L A_L \cdot \left(f_G + \frac{A_G}{e} \right). \tag{4.8}$$

4.5 Data Requirements

The determination of the maximum load that can be supplied without failure at each load bus can be accomplished in several ways depending upon the degree of accuracy required. The most exact, though certainly not the quickest, way is by conventional load-flow analysis. The data required in this case are as follows:
1. All the data required for load-flow analysis.
2. VAR limits at generating buses and voltage limits at all buses.
3. Maximum current-carrying capability of lines and transformers.
4. The distribution of the system load at various buses and the normalized load-duration curves.
5. The generation schedule at the increasing load levels being considered in the method. This is approximated to a straight-line variation.
6. Failure rate and repair rate of the generating units, lines, and transformers in the system.

Data Requirements

The load variation at each bus can be represented by a suitable probability distribution. The individual bus load levels used in each load-flow analysis correspond to a specified probability of being exceeded.

A load bus is assumed to be failed under the following conditions:
1. The voltage at the bus is less than a specified minimum value (not meeting the quality standards at the bus).
2. A line or transformer supplying power to the bus is overloaded.
3. The generating capacity required to meet the total load exceeds the available capacity. (This includes transmission losses.)

This list does not of course include all the possible modes of load-point failure. This is probably the most difficult part of any reliability analysis, that is, the recognition and classification of how a component and/or system can fail. Any other known or desired classifications can be included and treated in the same general way. The literature describes[8] a digital computer program that has been developed to create the possible outage conditions and to perform a load-flow analysis for each outage condition and specified load level. Under each outage condition B_j, if a load bus fails at any of the increasing load levels, P_L in Equation 4.2 is taken as the average of the probability value of the load level at which the load bus failed and the previous lower level. There is zero if the load bus does not fail even at the peak load level, and P_G in Equation 4.2 is the cumulative probability of the capacity outage in the system exceeding the reserve.

If a load bus is isolated from the network due to some outage condition, P_L is equal to 1 for that bus. If a generating bus is isolated, the capacity-outage probability table for the system is modified by removing[2] the generating units at the isolated bus before using it for computing P_G. If there is a load at the isolated generating bus, the contribution to the bus risk is computed by combining the load and the generating-capacity models at the bus as in a normal loss-of-load probability study. If the isolated generating bus is the swing bus, another bus is selected for this purpose.

If the load-flow solution does not converge under an outage condition within a specified maximum number of iterations, the system is divided into subsystems, with each generating bus supplying one or more neighboring loads. The division into subsystems should agree with accepted system practice, and therefore the splitting logic should be predetermined and inserted in the reliability-study phase. The risk contribution is then calculated for each bus using the capacity-outage probability table at the generating bus and a combined load-duration curve for the loads being supplied. If a line or transformer is overloaded at any load level, it is assumed to be tripped by the protective equipment and therefore removed from the network.

The probability steps to increase the load level are specified in the initial data. The maximum load level for any given outage condition can be determined more accurately by increasing the number of steps; however, the computation time also increases. It was observed that sufficiently accurate results can be obtained by considering up to a maximum of two simultaneous independent outages. Outages of higher order contribute negligible quantities to the total risk.

The risk evaluation is performed only if the base case load-flow solution with all the system components in service is satisfactory. The system load-flow solution was obtained using the conventional Gauss-Seidel method, but any solution technique can be used.

The system loads can be represented by either constant-power or constant-admittance models. The loads can also be represented by constant-power models, and if the voltage falls below a specified minimum, the model is modified to a constant-admittance representation. A constant-admittance representation increases the diagonal dominance of the nodal-admittance matrix and results in a significant saving in solution time.

From a practical point of view, the constant-admittance representation should be quite adequate as it is not necessary to obtain an exact voltage solution at a particular bus, only to find out if the voltage is above or below a set value. Some error will exist if the calculated voltage is just at the set point. If the load is supplied from the major transmission bus by on-load tap-changing transformers then it may be necessary to utilize a constant-power representation. In the study shown in this text, there is a considerable difference in the results obtained using the two-load representations. This is in part due to the small system example.

The system shown in Figure 4.6 was studied to compare the two approaches given by Equations 4.2 and 4.3 and also to investigate the effect of the number of load probability steps, the load representation, the bus-voltage limits, and the addition of selected lines. The data re-

Data Requirements 117

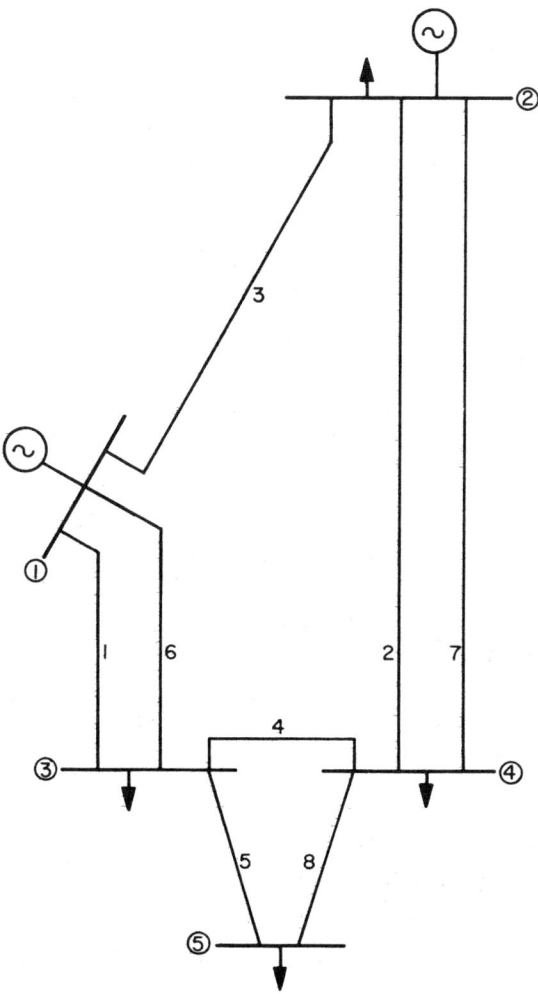

Figure 4.6 System studied for composite system reliability evaluation.

Applications to Bulk Power-Supply Systems   118

quirements for a study of this type have been previously noted. The data required in the system of Figure 4.6 are shown in Tables 4.2, 4.3, and 4.4.

It has been previously noted that many organizations such as the IEEE and the EEI are collecting outage data suitable for use in probabilistic reliability studies. The failure rates used in Tables 4.2 and 4.4 were synthesized from appropriate component data in a manner similar to that used in Example 2.1 on page 15. The failure rate of the line itself is proportional to its length. The terminal elements must also be included in the equivalent line-element value. The equivalent line-failure rate, repair rate, and the resulting probability-of-line-failure value therefore include data on all the elements required for the line to be effective. These elements are essentially connected in series from a reliability point of view, regardless of their actual physical connection. The amount of component data and the calculations required to synthesize the line parameters depend primarily upon the terminating configuration.

The risk levels in terms of probability of failure and expected frequency of failure for loads at buses 2, 3, 4, and 5 are shown in Table 4.5. Case 1 of Table 4.5 can be considered as the base case. Equations 4.2 and 4.3 have been used and the loads represented as constant admittances. A maximum of two simultaneous independent outages were

Table 4.2 The Generation Data for the Reliability Evaluation of the System Shown in Figure 4.6

Base MVA = 100
Base kV   = 110

Bus No.	No. of Units	Capacity of Each Unit (MW)	Total Bus Capacity (MW)	Type of Units	Failure Rate/Unit (failures per yr)	Repair Rate/Unit (repairs per yr)	Probability of Outage
1	4	20	80	Thermal	1.1	73	0.015
2	7	5	130	Hydro	0.5	100	0.005
	1	15		Hydro	0.5	100	0.005
	4	20		Hydro	0.5	100	0.005

Swing bus: 1.
(If bus 1 isolated from the network due to an outage condition bus 2 is selected as swing bus.)

Data Requirements

Table 4.3 The Load Data for the Reliability Evaluation of the System Shown in Figure 4.6

Bus No.	Peak Load (MW)	Power Factor	Generation Allotted under Peak Load (MW)	VAR Limits (MVAR)	Voltage Limits (p.u.) Max.	Min.
1	0	—	Swing bus	−10 to +10	1.05	0.97
2	20	1.0	110	0 to 40	1.05	0.97
3	85	1.0	—	—	1.05	0.97
4	40	1.0	—	—	1.05	0.97
5	10	1.0	—	—	1.05	0.97

The peak-load probability distribution was represented by a straight line from the 100% to 40% peak-load points.

Load-probability steps: 1.00 0.80 0.60 0.40 0.20 0.0 — (5 steps)
1.00 0.90 0.80 0.70 0.60 0.50
0.40 0.30 0.20 0.10 0.0 — (10 steps)

Table 4.4 The Line Data for the Reliability Evaluation of the System Shown in Figure 4.6

Lines are assumed to be 795 ACSR 54/7.
Current-carrying capability = 900 A = 1.71 p.u.
Failure rate = 0.05 failure/yr-mi.
Expected repair duration = 10 h.

Line	Length (mi)	Impedance (p.u.)	Susceptance ($b/2$)	Failure Rate (failure per yr)	Probability of Failure
1, 6	30	$0.0342 + j0.1800$	0.0106	1.5	0.001713
2, 7	100	$0.1140 + j0.6000$	0.0352	5.0	0.005710
3	80	$0.0912 + j0.4800$	0.0282	4.0	0.004568
4, 5, 8	20	$0.0228 + j0.1200$	0.0071	1.0	0.001142

Tolerance for load flow solution = 0.0001.
Maximum no. of iterations for convergence = 200.
Maximum no. of simultaneous independent outages of any combination considered = 2.

Applications to Bulk Power-Supply Systems

considered for the system consisting of lines 1 through 6. The load model represented by a straight line from the 100% to the 40% load points was approximated by 10 equal probability steps. The minimum acceptable system voltage was 0.97 per unit and the maximum generation voltage 1.05 per unit.

The results of Case 2 were obtained using Equations 4.4 and 4.5. The risk at bus 2 was found to be extremely low using this approach. This is partially due to the use of a maximum of two simultaneous independent outages. The risk at bus 2 in Case 1 is basically a system-generation contribution rather than an actual failure at bus 2. This effect is eliminated in Case 2 in which the complete system capacity outage probability table is not used. Case 3 shows the increase in risk at buses 3 and 5 when the system load models are changed to a constant-power representation. The effect of modifying the minimum acceptable voltage levels is shown in Cases 4a through 4h and is plotted in Figure 4.7. The effect of removing a line and adding additional lines is shown in Cases 5, 6, and 7.

In Case 8 the generating capacity was assumed to be completely reliable. The risk at bus 2 is therefore zero in this case. The risk levels at other buses in the system are reasonably close to those obtained in Case 1. Case 9 shows the effect of increasing the system loads. If the increase in risk violates the system design criterion, then the system must be modified to reduce the risk to an acceptable value. This concept is the basis of the logical transmission-planning program described in detail in the literature.[9] The assumption of complete generation reliability may be necessary in certain cases to eliminate the effect of the generation risk on the total risk upon which logical transmission planning is based.

The results obtained by considering only single independent outages, as shown in Case 11, are very close to those obtained in Case 1. This approach results in a significant saving in computer solution time over that required in Case 1. The bus risk levels using this approach in a system designed for single-contingency outages would be approximately equal to the values obtained by considering the generation adequacy only.

Case 12 shows the effect on the calculated values of using a five-step representation of the load characteristics. The results are significantly different than those obtained in Case 1. A reduction in the number of

Data Requirements

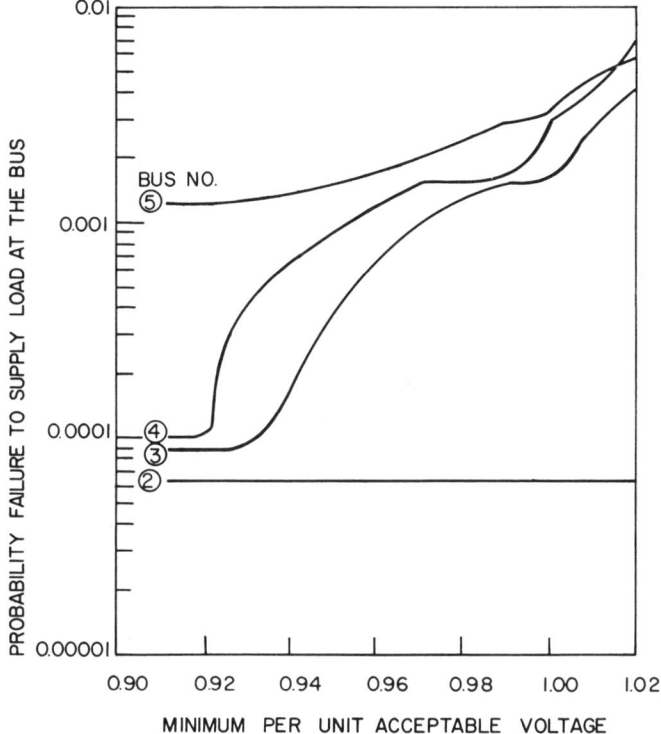

Figure 4.7 Effect of minimum acceptable bus voltage on the risk level.

steps may be required in a large system to reduce the computation time to an acceptable value.

Studies on a reduced configuration of the Saskatchewan Power Corporation in Canada indicated that there was relatively little difference between the results obtained using single independent outages and two simultaneous independent outages.[10] This may not be quite true in all systems and is dependent upon the degree of redundancy built into the system. The key word in this case is "independent." The probability of two or more simultaneous independent transmission outages is normally quite small and does not add substantially to the risk levels. If there is a degree of dependence associated with the outage of two or more facilities, then this must be included. The digital computer program described

Table 4.5 Risk Levels for Loads in the System Shown in Figure 4.6

Normal Case: Lines 1 through 6 in service; up to 2 simultaneous independent outages considered; load distribution approximated to 10 steps; loads represented as constant admittances; minimum acceptable voltage at all buses = 0.97 per unit; maximum allowed voltage at generating buses = 1.05 per unit; Equations 4.2 and 4.3 used.

Case	Changes from the Normal Case	Bus 2 Probability	Bus 2 Expected Frequency	Bus 3 Probability	Bus 3 Expected Frequency	Bus 4 Probability	Bus 4 Expected Frequency	Bus 5 Probability	Bus 5 Expected Frequency
1	Normal case	0.000062	0.0013	0.000937	0.8060	0.001729	1.5147	0.002075	1.8335
2	Generating unit outages considered individually (Equations 4.4 and 4.5)	0.000000	0.0000	0.000866	0.8114	0.001654	1.5315	0.001876	1.7335
3	Load representation as constant power	0.000063	0.0025	0.001500	1.3067	0.001730	1.5156	0.002638	2.3332
4a	Minimum acceptable voltage = 1.02 p.u.	0.000062	0.0016	0.004150	3.6707	0.006236	5.5331	0.005622	4.9929
4b	Minimum acceptable voltage = 1.01 p.u.	0.000062	0.0015	0.002910	2.5652	0.003872	3.472	0.004047	3.5900
4c	Minimum acceptable voltage = 1.00 p.u.	0.000062	0.0015	0.001727	1.5111	0.002747	2.439	0.002864	2.5364
4d	Minimum acceptable voltage = 0.99 p.u.	0.000062	0.0013	0.001727	1.5096	0.001732	1.5195	0.002864	2.5361

Data Requirements

4e	Minimum acceptable voltage = 0.95 p.u.	0.000062	0.0013	0.000376	0.3082	0.000942	0.8142	0.001514	1.3364
4f	Minimum acceptable voltage = 0.93 p.u.	0.000062	0.0013	0.000092	0.0544	0.000377	0.3112	0.001231	1.0825
4g	Minimum acceptable voltage = 0.91 p.u.	0.000062	0.0013	0.000090	0.0505	0.000097	0.0626	0.001229	1.0787
4h	Minimum acceptable voltage = 0.91 p.u.	0.000062	0.0013	0.000090	0.0505	0.000095	0.0587	0.001228	1.0787
5	Line 6 removed	0.000064	0.0043	0.005108	4.5116	0.005561	4.9148	0.006242	5.5278
6	Line 7 added	0.000061	0.0022	0.000069	0.0155	0.000069	0.0169	0.001207	1.0552
7	Lines 7 and 8 added	0.000061	0.0024	0.000067	0.0126	0.000067	0.0121	0.000068	0.0149
8	Generating capacity 100% reliable	0.0	0.0	0.000875	0.8040	0.001667	1.5128	0.002013	1.8315
9	Same as (8) but all loads increased by 10%	0.0	0.0	0.001439	1.3070	0.001669	1.5164	0.002576	2.3334
10	Same as (9) but line 8 added	0.0	0.0	0.000877	0.8095	0.001440	1.3118	0.000881	0.8165
11	Single outages only considered	0.000062	0.0013	0.000903	0.7470	0.001689	1.4432	0.002025	1.7458
12	Load distribution approximated to 5 steps	0.000062	0.0013	0.000657	0.5586	0.002233	1.9598	0.001795	1.5856

in Reference 2 removes any component that becomes overloaded due to the loss of another component. In this way, cascading-type situations can be recognized. This is essentially a steady-state solution and does not include transient disturbances and large swings in generation output. It is also possible to calculate the probability of each combination of independent outages prior to proceeding with a load-point study. If the probability is below a predetermined level, then this combination is discarded and the next combination examined. In this way, system analysis is performed only in those situations that make a reasonable contribution to the total load-point risk indices. Maintenance requirements can be treated by physically removing the component from the system and repeating the analysis. The load-model representation, however, should be compatible with the period of maintenance action, and in long-range transmission planning it may require extensive solution time to include the maintenance aspects. The effect of scheduled outages on the risk levels can be included by evaluating the risk levels under each scheduled outage and determining the annual sum of the risk levels weighted by the corresponding outage durations.

The reliability evaluation of single systems has shown that the risk levels calculated by the two different methods of considering generation inadequacy are not very different. The risk levels obtained by considering only single outages were found to be reasonably close to those obtained by considering two simultaneous independent outages. Considerable saving can be obtained in large systems using single outages only and also by reducing the load probability steps to an acceptable number. The effect on the bus risk levels of acceptable voltage limits, the load representation, and the removal or addition of lines are shown in Table 4.5. The computation time required to obtain the reliability indices for each bus in a system is a function of the technique used. The variation in calculated risk with selected changes in the approach are shown in Table 4.5. The use of any one of these techniques will lead to more consistent transmission planning than that obtained by the use of rule-of-thumb reliability criteria.

4.6 Transmission Planning

The economic examination of alternative generation and transmission expansion plans is much more meaningful when the reliability of each scheme is included in the analysis. Two plans can be compared economically, but if they have widely different reliabilities the analysis simply indicates which costs the most, not which is the best for the system under study. The concepts outlined in this chapter for composite system reliability evaluation have been utilized in sequential expansion planning.[9,10]

The cost of transmission improvements increases when a higher reliability level is demanded. The application of a quantitative reliability criterion facilitates optimum utilization of the capital available for transmission improvements. Expansion patterns with and without a quantitative criterion cannot be simply compared by advancing the installation of some transmission facilities by a fixed period to include the reliability requirement. A fixed contingency criterion without using the associated probability values can result in higher investment than required at some locations, particularly if a lower reliability level can be tolerated at these points. The selection of an acceptable risk level at each load point in the system is a planning decision, as is normally the selection of an adequate generating-capacity criterion. The utilization of individual load-point reliability indices determined by a consistent approach permits the planning engineer to include the costs of maintaining these levels in alternate planning proposals.

References

1. V. M. Cook, C. D. Galloway, M. J. Steinberg, and A. J. Wood, "Determination of Reserve Requirements of Two Interconnected Systems," *IEEE, Transactions on Power Apparatus and Systems*, vol. 82, 1963, pp. 18–33.

2. R. Billinton, *Power System Reliability Evaluation*. New York: Gordon and Breach Science Publishers, 1970, p. 299.

3. R. J. Ringlee and A. J. Wood, "Frequency and Duration Methods for Power System Reliability Calculations, Part V—Models for Delays in Unit Installations and Two Interconnected Systems," *IEEE, Transactions on Power Apparatus and Systems*, vol. 90, 1971, pp. 79–88.

4. R. Billinton and C. Singh, "Generating Capacity Reliability Evaluation in Interconnected System Using a Frequency and Duration Approach—Part I—Mathematical Analysis," *IEEE, Transactions on Power Apparatus and Systems*, vol. 90, 1971, pp. 1646–1654.

5. R. Billinton and C. Singh, "Generating Capacity Reliability Evaluation in Interconnected System Using a Frequency and Duration Approach—Part II—

System Applications," *IEEE, Transactions on Power Apparatus and Systems*, vol. 90, 1971, pp. 1654–1664.

6. R. Billinton, "Composite System Reliability Evaluation," *IEEE, Transactions on Power Apparatus and Systems*, vol. 88, 1969, pp. 276–280.

7. R. Billinton and M. P. Bhavaraju, "Transmission System Reliability Methods," IEEE Paper No. 71 TP 91-PWR, 1971 Winter Power Meeting.

8. R. Billinton and M. P. Bhavaraju, "Transmission Planning Using a Reliability Criterion—Part I—A Reliability Criterion," *IEEE, Transactions on Power Apparatus and Systems*, vol. 89, 1970, pp. 28–34.

9. R. Billinton and M. P. Bhavaraju, "Transmission Planning Using a Reliability Criterion—Part II—Transmission Planning," *IEEE, Transactions on Power Apparatus and Systems*, vol. 90, 1971, pp. 70–78.

10. R. Billinton and M. P. Bhavaraju, "Transmission Planning Using a Quantitative Reliability Criterion," *IEEE, PICA Conference Proceedings*, 1969, pp. 537–546.

5 GENERATION-SYSTEM OPERATION

5.1 Introduction

The determination of an acceptable level of reliability in the system operating area is an extremely difficult problem. One aspect of this problem is the determination of the operating capacity margin that will reasonably ensure sufficient capacity to meet the load. This capacity must be capable of satisfying unforeseen changes in system load and withstanding possible forced outage of some of the operating capacity.

These two aspects have been incorporated in a single risk index,[1] the probability of just carrying or failing to carry the system load. The determination of this risk index has been described in considerable detail in several publications.[2] Once established, this index provides the basic reference level upon which short-term economic scheduling can be evaluated.

Consider a single generating unit that can exist in one of two states, available and forced out of service. The state transition diagram is shown in Figure 5.1.

If the unit is in the "up" state at time $t = 0$, then[3]

$$P_{up} = \frac{\mu}{\lambda + \mu} + \frac{\lambda}{\lambda + \mu} e^{-(\lambda + \mu)t}$$

and

$$P_{down} = \frac{\lambda}{\lambda + \mu} - \frac{\lambda}{\lambda + \mu} e^{-(\lambda + \mu)t}. \qquad (5.1)$$

If the average repair time of the unit is much longer than the period of interest t, the probability of the unit failing during a short interval of time t is

$$P_{down} = 1 - e^{-\lambda t}.$$

If $\lambda t \ll 1$,

$$P_{down} \approx \lambda t = \text{O.R.R. (Outage Replacement Rate)}. \qquad (5.2)$$

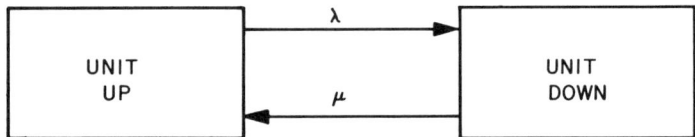

Figure 5.1 Two-state model.
λ = unit failure rate
μ = unit repair rate

This is the basic statistic used in many operating-capacity reliability studies, that is, the probability of the unit failing during the interval t when no changes can be made in the capacity available to carry the load. Using this statistic, it is relatively simple to develop a capacity-outage probability table for the capacity in or scheduled to be in operation. The cumulative probability of a capacity outage equal to the operating reserve is the probability of just carrying or failing to carry the system load. This can be easily illustrated using a simple example. Consider a hypothetical system having operating capacity equal to 250 MW consisting of the units listed in Table 5.1.

As indicated, a 4-hour lead time (i.e., $t = 4$ in Equation 5.2) was used. These risk indices at different load levels are given in Table 5.2.

At a forecast load of 220 MW, the probability of just carrying or failing to carry the system load is 0.00190151.

The uncertainty that always exists in load forecasting can be easily added to the analysis provided the distribution of load-forecast un-

Table 5.1 Hypothetical 250-MW Generating System

No. of Units	Type	Capacity (MW)	Failure Rate (failures per yr)	O.R.R. (4-h lead time)
4	Hydro	10.0	1	0.00045662
3	Hydro	15.0	1	0.00045662
5	Thermal	15.0	3	0.00136986
3	Thermal	20.0	3	0.00136986
1	Thermal	30.0	4	0.00182648

Introduction

Table 5.2 Risk Indices for 250-MW System

Capacity Out MW	Capacity In MW	Cumulative Probability
0	250	1.00000000
5	245	0.01586441
10	240	0.01586441
15	235	0.01406609
20	230	0.00596747
25	225	0.00191631
30	220	0.00190151
35	215	0.00006492
40	210	0.00003159
45	205	0.00002268
50	200	0.00000775
55	195	0.00000021
60	190	0.00000014

certainty is known. Assume in the previous example that the distribution shown in Figure 5.2 applies.

The risk for this load forecast is obtained as follows:

$$R = 0.2(0.00003159) + 0.6(0.00190151) + 0.2(00596747)$$

$$= 0.002340718. \quad (5.3)$$

The risk is in general higher in the case of an uncertain load. In the previous case with no uncertainty, if the risk level at the forecast load of 220 MW is higher than the predetermined acceptable level, then additional capacity must be scheduled for the period in question. The assumption is made in the utilization of a 4-hour lead time that there is additional capacity but it would be 4 hours before it could be made available to carry load.

The unit model shown in Figure 5.1 assumes that the unit can reside in one of two states "up" and "down". Markov process concepts can be utilized to determine the transient probabilities associated with a generating unit containing an additional derated state.[4] The general

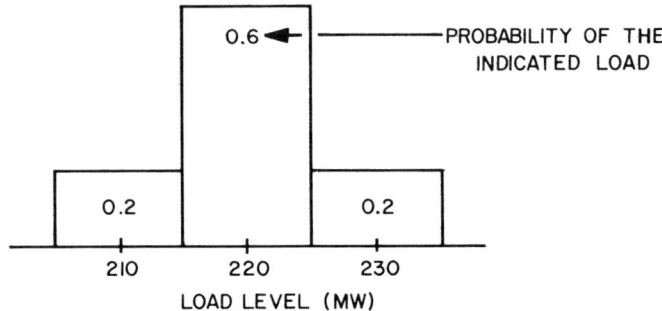

Figure 5.2 Distribution of load-forecast uncertainty.

expressions are rather complicated, and it is difficult to extend this approach to include units with more than one derated state. A state-space model has been proposed that assumes that the lead times are small relative to the unit repair times, and therefore it is reasonable to assume that no repairs will be made during this lead time.[4] This method, though approximate, can be applied to units with several derated states.

The Markov model for a unit with many possible derated states can be determined from the unit and auxiliary system design. A systematic approach has been proposed for the development of a multiderated state model and its associated transfer rates.[5] Methods have also been described for the calculation of the transient probabilities of being in each of the states. These probabilities can be designated as outage-replacement rates[3] or probabilities, that is, the probabilities of having the various levels of capacity forced out of service at some time t given that the unit was in a known condition at time $t = 0$. The matrix multiplication approach suggested in this paper[5] is also approximate but results in a virtually negligible error. The difference between the approximate and exact solutions is quite small under practical circumstances.

A more general formulation of the operating-capacity problem has been presented in terms of a breach of security.[6] This has been defined as follows:

an inadequacy of spinning generation capacity, unacceptable low voltage somewhere in the system, transmission line or other equipment overload, loss of system stability or other intolerable operating condition.

The mathematical formulation of the security function $s(t)$ or prob-

Derated State Concepts

ability of a breach of system security in the near future given a known operating condition at the time calculations are made is as follows:[6]

$$s(t) = \sum_i P_i(t) \cdot Q_i(t), \tag{5.4}$$

where

$P_i(t)$ = probability that the system is in state i at time t

and

$Q_i(t)$ = probability that state i constitutes a breach of security at time t.

Equation 5.4 can be seen to be a particular application of Equation 4.1 on page 109, which gives the probability of an event occurring when this event is dependent upon a group of mutually exclusive events. Equation 4.4 on page 113 is a similar application but with a slightly different physical interpretation. The form of security function shown in Equation 5.4 has been extended in the literature[7] to include both steady-state and transient breaches of security.

The probability of a unit residing in either an available or unavailable state as a function of time was given in Equation 5.1, where the unit was assumed to be available at time $t = 0$. These equations were simplified in Equation 5.2 by assuming that the likelihood of repair is negligible in the short period of interest. A capacity-outage probability table as shown in Table 5.2 represents an enumeration of the system capacity state probabilities at the lead time of 4 hours. The intersection of this capacity model and the load-forecast uncertainty model through convolution of the appropriate probabilities provides the risk index as calculated in Equation 5.3. This risk index is the first portion of the security function defined earlier.[6] The composite system reliability concepts or simply a conditional probability approach as illustrated in Chapter 4 can be utilized to include the remaining elements of the security-function definition.

5.2 Derated State Concepts

Economics and technological development have dictated the utilization of large-capacity generating units in recent years. Present-day (circa

1973) steam power plants can have single-unit ratings on the order of 1000 MW. Such a size may require multiple boilers and turbines and a large quantity of auxiliary equipment; therefore, in general, steam power plants of high capacity are subjected to a large number of possible derated capacity states. The unit may not be able to develop full rated capacity due to outages of boilers and auxiliaries such as pulverizers, induction draft fans, circulating water pumps, and so on; fuel quality limitations may also cause excessive slagging that can reduce the maximum unit output by 5% to 10%. Environmental conditions can also be responsible for a reduction in the plant capacity due to altering the fuel quality, pollution, or by causing other limitations on the equipment. Some of these factors can be closely predicted for the next 24 hours, and if the system has sufficient installed capacity, predictable deratings do not cause serious operational difficulties. Deratings normally range from 5% to 50% of the unit capacity, and deratings beyond 50% of the unit capacity are rare. For a medium- to large-sized steam power plant, some of the important auxiliaries are the following:
1. Forced draft fans and induction draft fans
2. Gas recirculation fan
3. Primary air fan for the pulverizer
4. Circulating water pumps
5. Boiler feed booster pumps
6. Condensate pump
7. Pulverizer drives
8. Soot blower air compressor
9. Ash removal pumps

In a typical installation some of these auxiliaries (i.e., numbers 2, 6, 8, 9) may have two fully redundant sets of equipment, and if one fails, the other takes over operation very quickly. As soon as it is feasible, the failed auxiliary or the failed auxiliary drive is repaired. During the short period when one auxiliary is failed, the probability of the other failing is quite remote. Therefore, such auxiliaries do not normally cause forced deratings.

Certain other auxiliaries such as forced draft fans, induction draft fans, pulverizers, and circulating water pumps may not be 100% redundant. If a boiler loses one forced draft fan out of two, its maximum output will be reduced by a known factor. Insufficient circulating cooling

Derated State Concepts

water due to the failure of a circulating water pump may cause the maximum output to be reduced. Certain other outages may cause a total boiler outage. The multitude of components and auxiliaries can therefore produce many partial outages.

5.2.1 Detailed State-Space Diagram of a Large Generating Unit
Consider a 200-MW thermal unit having the following components causing dominant outages:
3 feed water pumps
3 circulating water pumps
2 induction draft fans
2 forced draft fans
5 pulverizers
1 boiler
1 turbine

There are 15 main components causing dominant outages, giving rise to $2^{15} - 1$ possible outage conditions. Assuming the transitions to different states to be a stationary Markov process, the state-space diagram becomes unwieldy. This diagram can be simplified by omitting the impossible states and merging the identical component outages. It has been noted in the preceding section that during the lead time, the probability of failure of more than one auxiliary component of the same type is extremely small and therefore can be neglected in most practical cases. Only the identical component outage states are lumped together at this state. A recursive approach can be used to obtain the equivalent transfer rates for any new state. The availability of the new state is equal to the sum of the availabilities of the merged states. The frequency of encounters in the new state must be equal to those in the original system. Identical component outage states in general cannot communicate.

Consider for example that the unit has five similar pulverizers and a single failure causes a 20% reduction in the maximum capability. Assume that
A_P = probability of outage of each pulverizer,
A = probability that the unit is in the up state,
λ_P = failure rate of each pulverizer,
μ_P = repair rate of each pulverizer,
λ'_P = failure rate of the group of merged pulverizers,

Generation-System Operation

μ'_P = repair rate of the group of merged pulverizers,
and
A'_P = probability of outage of the group of merged pulverizers.
Now,

$$A'_P = 5 \cdot A_P (1 - A_P)^4.$$

Equating the rates of transfer from the full-capacity state to the one-pulverizer-failed state gives

$$\lambda'_P = 5\lambda_P.$$

Equating the rates of transfer from the one-pulverizer-failed state to the full-capacity state gives

$$\mu'_P = \mu_P.$$

The state-space diagram for this case is given in Figure 5.3. In this diagram only the transitions between the up state and the various lower-capacity states are shown. It has been assumed that no more than one auxiliary of the same kind fails within the lead time. If a large number of components of the same type are present, the possible outage of more than one component will have to be considered. Other cases of transitions between the derated states and the failed states may also be considered without any difficulty.

Deratings due to fuel quality restrictions, such as sodium in the coal causing excessive slagging, or to environmental effects are not represented in Figure 5.3. The failure of the hydrogen cooling may also cause the unit to lose as much as one-third to one-half of the rated capacity. A more exact state-space diagram for a unit, which includes these effects, is given in Reference 5.

5.2.2 Equivalent State-Space Model of a Large Generating Unit
Even after merging the identical component outage states, the state-space model is generally still too complicated and difficult to handle.

Derated State Concepts   135

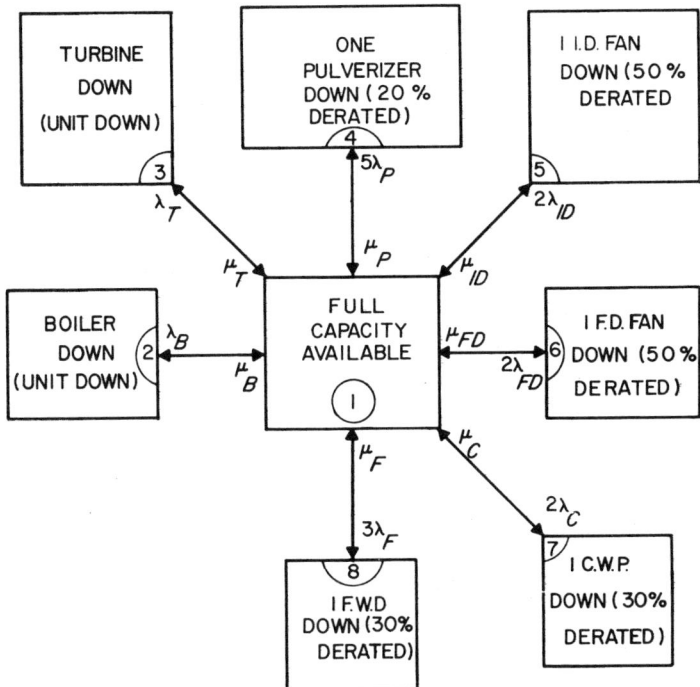

Figure 5.3 Generating unit state-space diagram.

It is desirable to reduce it still further without sacrificing accuracy. This may be accomplished by merging the equal capacity states.

The equivalent state-space diagram of a unit is given in Figure 5.4. The method should prove useful for evaluating the large system generating units that are now being commissioned; these have components and auxiliaries for which the outage statistics are well established. The method can be applied to multiple boiler units and the exact and equivalent state-space diagrams obtained.

5.2.3 Calculation of the Transition Probabilities

The differential matrix shown in Table 5.3 represents a set of simultaneous linear differential equations. The order of the equations is equal to the total number of equivalent states; that is, six simultaneous linear

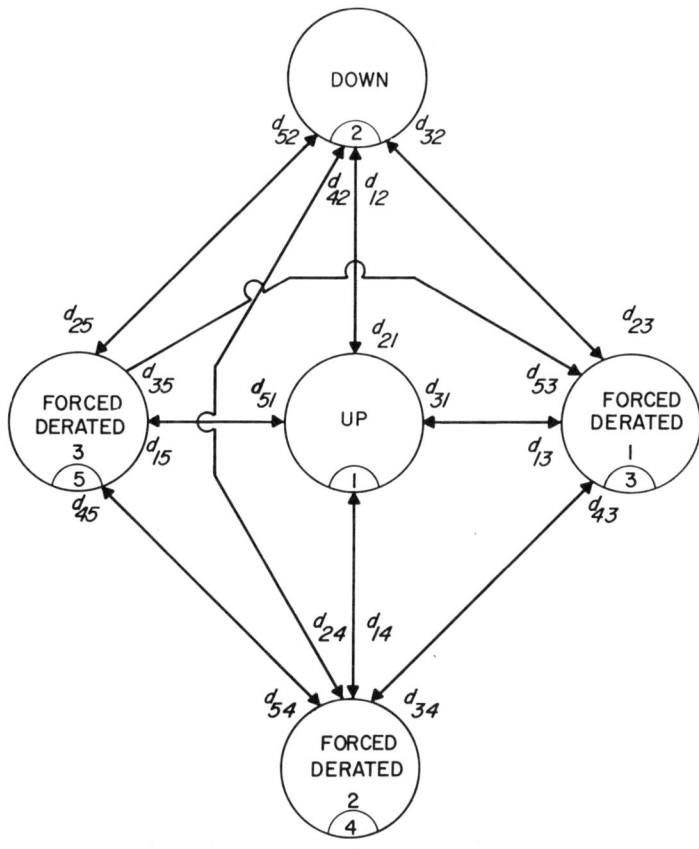

Figure 5.4 State-space representation of a unit model with three derated states.

differential equations will result from a unit having four derated capacity states. It is difficult to find a general time-dependent solution for a system having more than three states. Numerical methods can be applied to get the time-dependent solution of these equations, but they are rather time consuming, especially when there are many derated states. Other methods are available, however, that are both easy to apply and very efficient in terms of computer time. The approach presented in Reference 5 considers the continuous Markov process as a discrete one; that is, transitions from one state to another take place in discrete periodic steps. The development therefore proceeds as follows.

Table 5.3 Differential Probability Matrix for a Unit with Three Derated States

Referring to Figure 5.4, 1 is the up state, 2 is the down state, 3, 4, and 5 are the forced derated states.

$[P'(t)] = [A] \cdot [P(t)]$

Where $[A]$ is the following matrix:

$$\begin{bmatrix} -(d_{12}+d_{13}+d_{14}+d_{15}) & d_{21} & d_{31} & d_{41} & d_{51} \\ d_{12} & -(d_{21}+d_{23}+d_{24}+d_{25}) & d_{32} & d_{42} & d_{52} \\ d_{13} & d_{23} & -(d_{31}+d_{32}+d_{34}+d_{35}) & d_{43} & d_{53} \\ d_{14} & d_{24} & d_{34} & -(d_{41}+d_{42}+d_{43}+d_{45}) & d_{54} \\ d_{15} & d_{25} & d_{35} & d_{45} & -(d_{51}+d_{52}+d_{53}+d_{54}) \end{bmatrix}$$

Generation-System Operation

Consider a homogeneous Markov chain of the first order with transition probabilities:

$$P_{ij} = P\{\varepsilon_n = j \mid \varepsilon_{n-1} = i, \quad i,j = 1,2,\ldots\}$$

(probability of the process being in the state j at the nth interval, given that the process is in the state i at the preceding interval). Also

$$P(2) = P(1)\Pi,$$

where $P(1)$ and $P(2)$ are row vectors indicating the state probabilities after 1 and 2 steps, respectively, and Π is the following matrix:

$$\Pi = \begin{bmatrix} p_{11} & p_{12} & p_{13} & - & - \\ p_{21} & p_{22} & p_{23} & - & - \\ p_{31} & p_{32} & p_{33} & - & - \\ - & - & - & - & - \\ - & - & - & - & - \end{bmatrix}.$$

This matrix Π is square, its elements are nonnegative, and the rows add up to 1. If the process is initially in state i,

$$P_i(0) = 1.0$$

and

$$P_j(0) = \ldots = 0 \ (j \neq i).$$

The state probabilities $P_j(n)$ are calculated as follows:

$$P_j(n) = \sum_i P_i(0) p_{ij}^{(n)}.$$

Since the transition probabilities are already in matrix form, $p_{ij}^{(n)}$ will be the (i,j)th element of the nth power of Π. Therefore if Π is the stochastic transitional probability matrix for a 1-hour period, the transition probabilities for a 4-hour lead time can be obtained by multiplying the matrix by itself four times. The accuracy can be further improved by

Derated State Concepts

setting up the matrix for a half-hour interval and multiplying it by itself eight times. It was found that only a small error is introduced by using the matrix on a 1-hour basis. The results obtained by this method[3] have been compared with those obtained by the solution of the simultaneous linear differential equations using the Runge-Kutta method.

5.2.4 Effects of Including Forced Derated States

When the thermal units in a system are represented by their equivalent derated models, the risk levels obtained for the same loads at the given operating capacity are generally lower than those obtained without considering the deratings. Large thermal units show a multitude of deratings, and the outage probability assigned to the exact outage of full capacity of the unit is lower in this case. The correct inclusion of the derated states into the capacity model may result in either an increase or decrease in the risk level depending upon how the derated states had been previously incorporated. Consider the simple hypothetical system shown earlier in this chapter, and assume that each thermal unit is represented by the derated models shown in Table 5.4. Each state is represented by the transient probability of a 4-hour lead time.

Table 5.4 Derated Models for Thermal Units

Capacity Out (MW)	Probability
15-MW Unit	
0.0	0.99863014
5.0	0.00091324
15.0	0.00045662
20-MW Unit	
0.0	0.99863014
5.0	0.00091324
20.0	0.00045662
30-MW Unit	
0.0	0.99817352
5.0	0.00091324
15.0	0.00045662
30.0	0.00045662

The results, including those presented earlier, are shown in Table 5.5. It can be seen from Table 5.5 that the risk at a load of 220 MW is 0.00092952 as compared to 0.00190151 when deratings were considered as total outages.

Any analytical technique used to predict system or subsystem reliability should respond to the factors that do actually influence the reliability of the system. The models used to represent the system elements, in this case the operating generating units, should therefore be as complete as possible.[5] The addition of a generating unit with several capacity states into a previously computed capacity-outage probability table does not present any difficulties and is a simple extension of the usual two-state unit-addition method.[3] The effects of including derated levels in a spinning-reserve study will vary in each system. They will also depend upon how the derated levels were previously considered in regard to grouping at the full- and zero-capacity points. In many units,

Table 5.5 System Capacity-Outage Availability Probability Table

Operating capacity = 250 MW

Capacity Out (MW)	Capacity Available (MW)	Cumulative Probability	
		Case a	Case b
0.0	250.0	1.00000000	1.00000000
5.0	245.0	0.01586441	0.01586441
10.0	240.0	0.01586441	0.00821432
15.0	235.0	0.01406609	0.00638967
20.0	230.0	0.00596747	0.00232675
25.0	225.0	0.00191631	0.00094628
30.0	220.0	0.00190151	0.00092952
35.0	215.0	0.00006492	0.00001915
40.0	210.0	0.00003159	0.00000694
45.0	205.0	0.00002268	0.00000461
50.0	200.0	0.00000775	0.00000129
55.0	195.0	0.00000021	0.00000003
60.0	190.0	0.00000014	0.00000001

Probabilities less than 10^{-8} have been neglected.
Case a: Deratings considered as full outages.
Case b: Deratings included.

particularly large thermal equipment, there are several easily recognizable derated levels that should be included in the system reliability model.

5.3 The Effect of Peaking Equipment on the Operating Reserve

The delay time associated with additional generation is quite variable and is dependent upon many factors. The most important, however, is the type of additional generation available. Gas turbine peaking units and hydraulic equipment can usually be made available in a relatively short time as compared to more conventional thermal or nuclear generating capacity. Pumped storage hydraulic units operating in the pumping mode have a different delay time associated with their load-carrying capability. This is also true for conventional thermal equipment operating in a hot reserve mode. All these factors should be considered in a spinning-reserve study, and the reliability method should respond to the factors that influence the actual reliability of the system.[8]

5.3.1 Area Risk Curves

The probability of finding a unit on outage at time t if the unit is available at time $t = 0$ is given by Equation 5.1. This equation is repeated here:

$$P_{\text{down}} = \frac{\lambda}{\lambda + \mu} - \frac{\lambda}{\lambda + \mu} e^{-(\lambda + \mu)t}. \tag{5.1}$$

At time $t = 0$, $P_{\text{down}} = 0$.

As time $t \to \infty$, $P_{\text{down}} = \dfrac{\lambda}{\lambda + \mu} = \dfrac{r}{m + r}$

$\phantom{\text{As time } t \to \infty, P_{\text{down}}} = $ Forced Outage Rate.

The probability of finding the unit in the down state can be illustrated by representing it as an area rather than a numerical value. In the single-unit binary-model case, the probability is the area under a curve of the form

$$F(R) = \lambda e^{-(\lambda + \mu)t}.$$

The area is given by

$$\int_0^T F(R)\, dT = \frac{\lambda}{\lambda + \mu} - \frac{\lambda}{\lambda + \mu} e^{-(\lambda + \mu)t}.$$

The area risk curve for the single unit is shown in Figure 5.5.

The probability of finding the unit out of service at time t is now represented pictorially by an area rather than a single numerical value. When $t = \infty$, the area under the curve is equal to $\lambda/(\lambda + \mu)$, the conventional steady-state forced-outage probability. It should be made quite clear that the use of the area concept is simply to give a more physical picture of the risk. This will become more evident later in the chapter when rapid-start generating units are considered. The areas are not required in order to compute numerically the total system risk as a function of time. The state probabilities are obtained directly from the appropriate unit expressions in the form of Equation 5.1.

In a spinning-reserve study it is often assumed that sufficient additional generation is always available and that it is only a matter of time before this capacity is placed in service. If this delay time is 2 hours, then the risk is given by the area *abcd* in Figure 5.5, for 4 hours, by the area *abef*. The entire area to $t = \infty$ is equal to the generating-unit forced outage rate.

In an actual system, a number of units would be operating with additional generation available after a finite delay time. As previously noted, this additional generation can take many forms. The delay time asso-

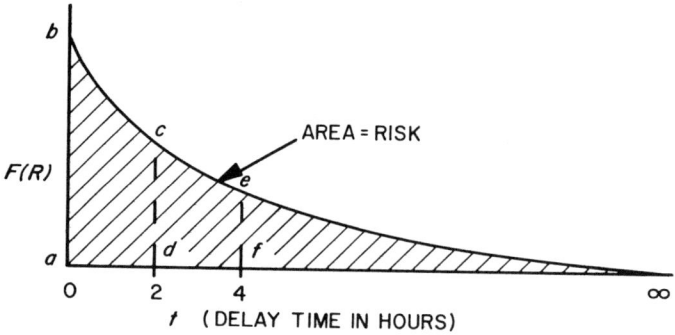

Figure 5.5 A single-unit area risk curve.

The Effect of Peaking Equipment on the Operating Reserve 143

ciated with conventional thermal generation may vary from 2 to 24 hours depending upon when the equipment last operated. Hydro and gas turbine units require a very short lead time to start, synchronize, and carry load. The loading rate in these cases is also quite different from that of large thermal equipment. The delay time of a conventional thermal unit can be reduced by maintaining the boiler in a banked state. Units in this condition are designated as hot reserve. The increase in cost must be balanced against the decrease in risk to the system, which is a function of the state of readiness of the units expressed in the form of delay time.

Consider a theoretical system with eight generating units for which the load requirements dictate that five units be operated. Assume that of the remaining three units, two can be placed in service after 4 hours and the third requires 8 hours' notice. The risk after 4 hours is of having less generation than load with the seven available units and after 8 hours, of having less generation than load with all the units. The area risk curve for this condition is shown in Figure 5.6.

The area under the curve from $t = 0$ to any particular point is dependent upon two factors, the available generation and the load level. The assumption has been made that the load is a constant value. If after the peak has been realized the system load level decreases, then the risk contribution decreases from that point on. The area risk curve may be truncated at that point. If the operating capacity is also reduced, then there will be a change in the capacity composition and an increase in risk. At any time $t = 0$, the question becomes, What is the risk associated with the known generation state and the expected load situation?

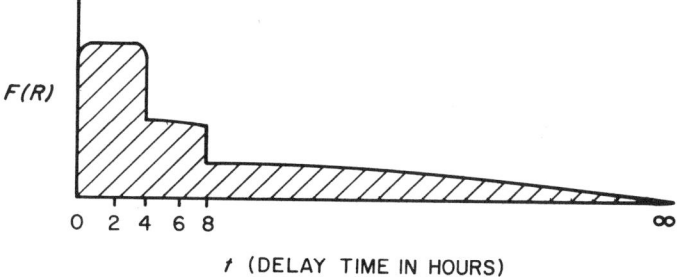

Figure 5.6 Area risk curves for five units operating out of eight.

Generation-System Operation   144

If this risk is too high, then the known generation state must be modified. Figures 5.7 and 5.8 show symbolic area risk curves for seven units operating out of eight and for all eight units operating, respectively.

The individual units may include one or more derated states, and load forecast uncertainty can be included in the analysis. It can clearly be seen from the area risk curves that the risk is a function of the delay time associated with the additional equipment. The risk is often negligible after additional generation becomes available. The effects of lead time and operating margin on the system risk are shown in Figure 5.9 for a 2900-MW hypothetical system.

The area risk curves in Figures 5.5 to 5.9 are shown to extend out to infinity. In actual practice they would only extend as far out as the load level applies. A change in load level would cause a resulting change in risk. This is the situation depicted by Equation 5.4 or by a time-dependent capacity-outage probability table intersecting with a chronological load model. As previously noted, the area risk concept is meant

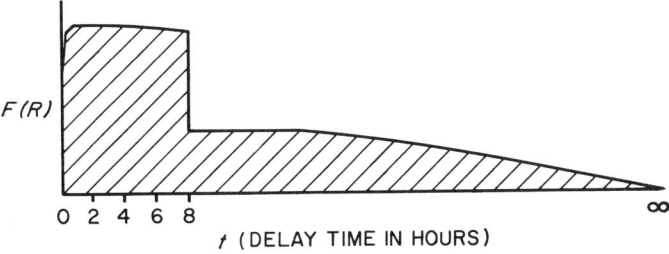

Figure 5.7 Area risk curves for seven units operating out of eight.

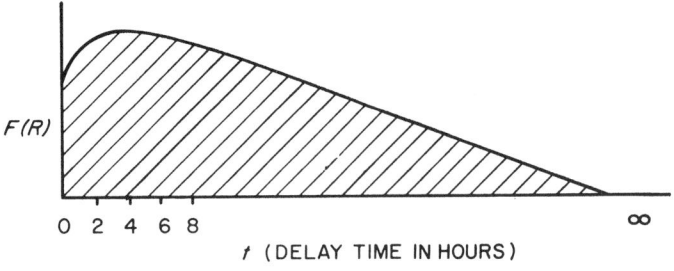

Figure 5.8 Area risk curves for all the units operating.

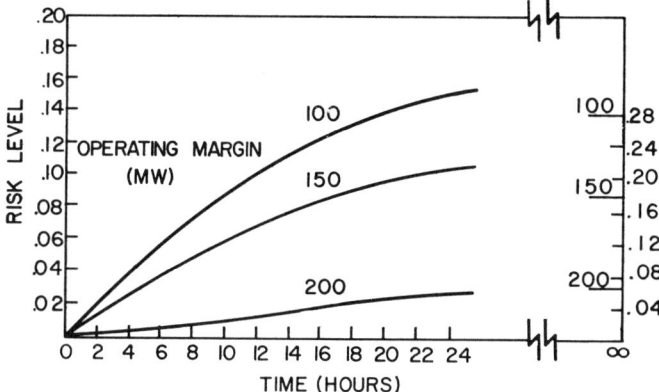

Figure 5.9 Variation of risk levels for different operating margins for increasing lead times in a hypothetical system.

only to illustrate the idea of a time-dependent risk. When a unit is scheduled for service and placed on line, the risk level immediately drops to a lower value, and in daily scheduling the risk will vary considerably with discrete unit additions and removals and continuous load-level variations. The object, of course, is to satisfy the system load requirement at the lowest total cost without violating a predetermined risk level.[6]

5.3.2 Rapid Start and Hot Reserve Units

The area risk curve concept can be used to illustrate the effect on the risk level of rapid start and hot reserve units. Consider a hypothetical system that contains some rapid start units that can be placed in service after a 5-minute delay. If the decision is made, the risk function $F(R)$ will decrease after 5 minutes to a new value dependent upon the quantity of rapid start capacity. This condition is illustrated in Figure 5.10, where it has been assumed that additional conventional generation is available after 4 hours.

The periods t_q and t_c are the times required to bring the rapid start and the conventional equipment into service. If the system also contains some hot reserve units with a delay time t_h taken to be in the order of 1 hour, the area risk curve takes the form shown in Figure 5.11.

In the limiting cases, an infinite amount of rapid start capacity would reduce the system delay time to 5 minutes, after which there would be

Generation-System Operation

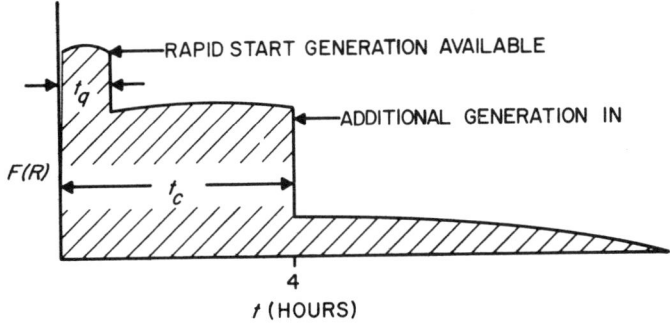

Figure 5.10 Effect of rapid start units on the risk level.

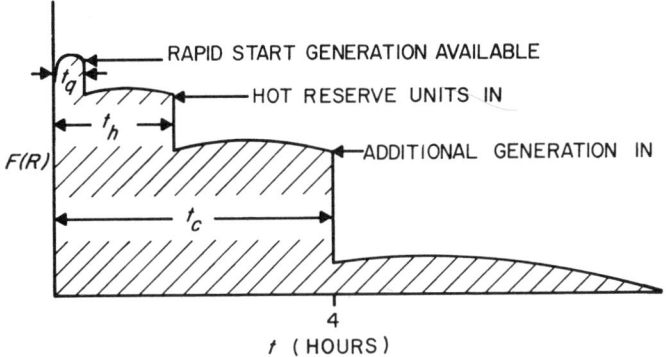

Figure 5.11 Effect of rapid start and hot reserve units on the risk level.

a negligible risk. In the case of an infinite amount of hot reserve this change would occur at the assumed 1-hour mark.

The generation available to carry load and exposed to failure in the rapid start and hot reserve case shown in Figure 5.11 is not constant. The system lead time can be divided into three parts and treated individually:

Period 1—Only the actual generation in service is available. This will extend over the period t_q. If no quick start unit is available this will extend up to t_h and up to t_c if there is no hot reserve.

Period 2—This will extend from t_q to t_h when both rapid start and hot reserve units are available.

Period 3—This will extend from t_h to t_c. The risk level will be quite small in this region if large quantities of rapid start and hot reserve exist; however, it may extend over a longer period.

It has been assumed that additional conventional generation becomes available after time t_c and that the risk function $F(R)$ becomes quite small. If this contribution to the area risk is not small, then an additional region must be added. In most cases, the required lead time is less than 12 hours, and the duration of the peak load is sufficiently small that extensive time spans are not required.

The risk-level contribution during the first time period can be computed using the standard procedure described in detail earlier. In the second time period, the possibility arises that the rapid start units may fail to start and therefore become unavailable to the system. This condition can be included using a standard conditional probability approach as follows:

probability of system failure

= (probability of the system generation and the quick start units just carrying or failing to meet system load | quick start units becoming available)

× (probability of quick start units becoming available)

+ (probability of the operating capacity just carrying or failing to carry the system load | quick start units not becoming available)

× (probability of quick start units not becoming available).

In the case of a unit on hot reserve, the probability of being unavailable is the probability of failure while banked plus the probability of failure to take up load. The hot reserve units are added to the capacity available in the third period. As indicated earlier, the area risk concept is for illustrative purposes only, the time-dependent capacity-outage probability tables are obtained from appropriate combinations of the unit states represented by Markov models.

The time-dependent probabilities for the time periods t_q, t_h, and t_c can be easily obtained.[5] The applicable model is used for each unit in operation, that is, thermal, hydraulic, rapid start, and hot reserve units, and so on.

5.3.3 Rapid Start and Hot Reserve Unit Models

A rapid start unit is normally required at relatively short notice, and at that time can therefore behave in two ways. It can fail to start, or it can go into operation. When it fails to start, the unit is considered to have failed and is then checked for faulty components, repaired, and either brought into service if required or placed back in the waiting for service mode. A Markov model of such a system is shown in Figure 5.12.

A model for peaking units has recently been introduced by Calsetta and his co-workers,[9] proposing a four-state model that takes into account the effect of reserve shutdown of a peaking unit. This model gives the conditional probability of a rapid start unit not being able to serve when called into service. The model considers that all repairs do not take place immediately. In the case of spinning-reserve studies, the probabilities of a unit not being able to start when it is called upon and of its moving to the failed state from the operation state during the lead time can be calculated from the model given in Figure 5.12. During the short lead time, the rapid start units if available are always considered for possible operation in the event of any outage or contingency. A

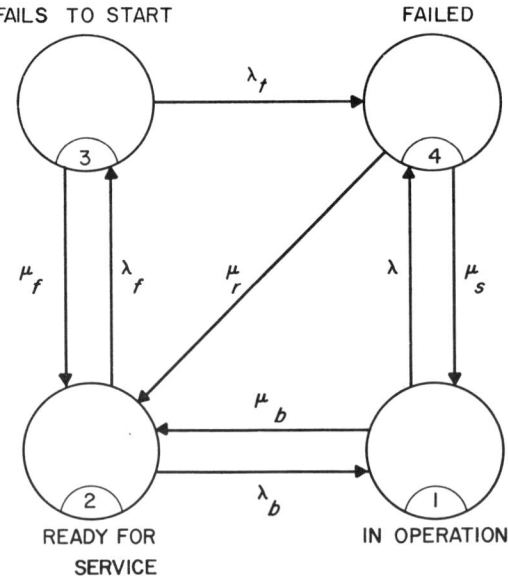

Figure 5.12 State-space diagram for a rapid start unit.

The Effect of Peaking Equipment on the Operating Reserve

large number of rapid start units are remotely controlled, and failures can take place due to the failure of the communication equipment. These repair rates will have very different average durations and distributions compared to the repair rates of the rapid start units themselves. The repair rate μ_f has been inserted to take this condition into account.

The departure or the transition rates from one state to the other can be obtained as follows:

N_S = number of times the unit fails to start,
N_F = number of times the unit is forced out,
N_T = number of successful transitions from the ready-for-service to the operation state,
N_B = number of times the unit is shut down though it is not actually forced out,
N_R = number of transitions from the failed to ready-for-service state,
N_K = number of times the unit is repaired back to service,
N_A = number of times the unit goes from the fails-to-start condition directly to the ready-for-service condition,

and

N_D = number of times the unit goes from the fails-to-start to the failed state.

It can be seen that

$$N_S + N_F = N_R + N_K + N_A.$$

The following definitions are made:
T_1 = total time spent in the operating state,
T_2 = total time spent in the ready-for-service state,
T_3 = total time spent in the failure-to-start state (it will be very small),
and
T_4 = total time spent in the failed state.

Therefore the transition rates are as follows:

$$\lambda_f = \frac{N_S}{T_2}, \qquad \lambda_b = \frac{N_T}{T_2},$$

$$\lambda_t = \frac{N_D}{T_3}, \qquad \lambda = \frac{N_F}{T_1},$$

Generation-System Operation

$$\mu_r = \frac{N_R}{T_4}, \qquad \mu_s = \frac{N_K}{T_4},$$

$$\mu_b = \frac{N_B}{T_1}, \qquad \mu_f = \frac{N_A}{T_3}.$$

The state-space diagram shown in Figure 5.12 gives the following differential probability matrix equation.

$$\begin{vmatrix} P_1'(t) \\ P_2'(t) \\ P_3'(t) \\ P_4'(t) \end{vmatrix} = \begin{matrix} 1 \\ 2 \\ 3 \\ 4 \end{matrix} \begin{vmatrix} -\lambda - \mu_b & \lambda_b & 0 & \mu_s \\ \mu_b & -\lambda_b - \lambda_f & \mu_f & \mu_r \\ 0 & \lambda_f & -\lambda_t - \mu_f & 0 \\ \lambda & 0 & \lambda_t & -\mu_s - \mu_r \end{vmatrix} \begin{vmatrix} P_1(t) \\ P_2(t) \\ P_3(t) \\ P_4(t) \end{vmatrix}$$

The time-dependent probabilities can be easily calculated from this array of differential equations.[3]

The state-space diagram for the units on hot reserve is basically the same as that shown previously for quick start units. The differences are that while the units are banked they can also fail and after a unit is repaired it will go back to the cold reserve state. Repairs on thermal units are normally made in the cold state and the unit cannot return to the hot reserve state immediately. A new state representing cold reserve must, therefore, be added. The state-space diagram is given in Figure 5.13, where it has been assumed that the unit will not be derated. If deratings can occur, the state-space diagram will be modified accordingly. In the case of a unit on hot reserve, the probability of residing in State 3 where the unit fails to take up load, will be extremely small. This state may be omitted entirely in many cases.

The additional transfer rates in this diagram over a rapid start unit require the following definitions:

T_5 = total time spent in the cold reserve state,
λ_r = transition rate from the hot reserve state to the failed state,
μ_r = transition rate from the failed state to the cold reserve state (repair rate),
μ_c = transition rate from the cold reserve state to the hot reserve state,
λ_c = transition rate from the hot reserve state to the cold reserve state,
μ_g = transition rate from the cold reserve state to the operation state,

The Effect of Peaking Equipment on the Operating Reserve 151

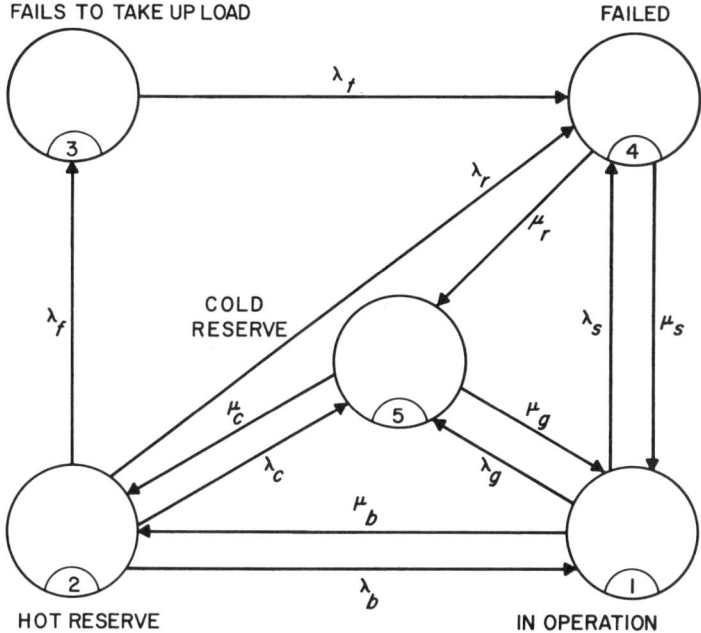

Figure 5.13 State-space diagram for a unit on hot reserve.

λ_g = transition rate from the operation state to the cold reserve state,

N_R = total number of transitions from the failed state to the cold reserve state,

N_F = number of times the unit is forced out,

N_P = number of times the unit moves from the cold reserve state to the hot reserve state,

N_Q = number of times the unit moves from the hot reserve state to the cold reserve state,

N_C = number of transitions from the cold reserve state to the operation state,

N_L = number of transitions from the operation state to the cold reserve state,

and

N_W = total number of transitions from the hot reserve state to the failed state.

Table 5.6 Differential Equations in Matrix Form for State-Space Diagram of Figure 5.13

		1	2	3	4	5	
$P'_1(t)$	1	$-\lambda_s - \mu_b - \mu_g$	λ_b	0	μ_s	μ_g	$P_1(t)$
$P'_2(t)$	2	μ_b	$-\lambda_b - \lambda_f - \lambda_r - \lambda_c$	0	0	μ_c	$P_2(t)$
$P'_3(t)$ =	3	0	λ_f	$-\lambda_t$	0	0	$P_3(t)$
$P'_4(t)$	4	λ_s	λ_r	λ_t	$-\mu_s - \mu_r$	0	$P_4(t)$
$P'_5(t)$	5	λ_g	λ_c	0	μ_r	$-\mu_c - \mu_g$	$P_5(t)$

The Effect of Peaking Equipment on the Operating Reserve

The additional transfer rates are as follows:

$$\lambda_s = \frac{N_F}{T_1}, \qquad \lambda_g = \frac{N_L}{T_1}, \qquad \mu_c = \frac{N_P}{T_5},$$

$$\mu_r = \frac{N_R}{T_4}, \qquad \lambda_c = \frac{N_Q}{T_2}, \qquad \lambda_r = \frac{N_W}{T_2},$$

$$\mu_g = \frac{N_C}{T_5}.$$

The differential equations in matrix form are given in Table 5.6.

Value of transfer rates λ_c and λ_g should be substituted as zero for spinning-reserve studies.

The time-dependent probabilities can be obtained from these simultaneous linear differential equations. The concepts developed in the previous sections were applied to the 2900-MW system consisting of 24 conventional thermal units described in Table 5.7. The transfer rates given in the table refer to the state transition diagram shown in Figure 5.14.

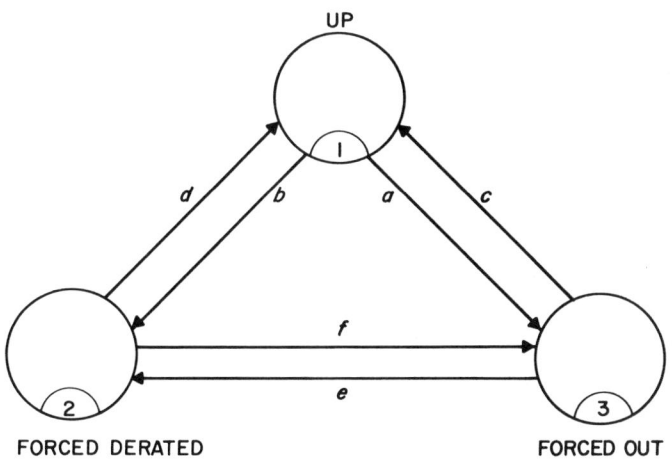

Figure 5.14 Biggerstaff and Jackson's[4] state-space model for a generating unit with one derated state.

Table 5.7 Transfer Rates for 2900-MW System

Number of Units	Size (MW)	Transfer Rates per Hour					
		a	b	c	d	e	f
15	100	0.0003	0.0010	0.0225	0.0350	0.0008	0.0004
8	150	0.0006	0.0050	0.0400	0.0400	0.0004	0.0004
1	200	0.0005	0.0002	0.0240	0.0430	0.0001	0.0001

The derated state shown in Figure 5.14 was assumed to be 80% of the full capacity rating.

The operating capacity in the system was assumed to be 2100 MW and the load 1900 MW. At the 2100-MW capacity level it was assumed that seven 100-MW units, eight 150-MW units, and one 200-MW unit were in operation.

The effects of different amounts of rapid start generation and units on hot reserve on the risk level were considered. The times to put the rapid start units and the hot reserve units into service were taken as 10 minutes and 1 hour, respectively. The capacities of the hot reserve units and the rapid start units together with the various transition rates used in this study are as follows:

Rapid Start Units
Size = 30 MW
Transfer rates/hour (refer to Figure 5.12)
$\lambda_f = 0.0001$ $\lambda_b = 0.0040$
$\lambda_t = 0.0250$ $\lambda = 0.0006$
$\mu_r = 0.0250$ $\mu_s = 0.0150$
$\mu_b = 0.0050$ $\mu_f = 0.0$

Probability of failure in State 2 ($P_{2-3} + P_{2-4}$) for
Lead time (2 hours) = 0.000203
Lead time (4 hours) = 0.000412
Lead time (6 hours) = 0.000627
Lead time (8 hours) = 0.000845
Lead time (12 hours) = 0.001290

Units on Hot Reserve
Capacity = 50 MW
Transfer rates/hour (refer to Figure 5.13)

The Effect of Peaking Equipment on the Operating Reserve

$\lambda_f = 0.00002$ $\quad\quad$ $\lambda_s = 0.00200$ $\quad\quad$ $\lambda_c = 0.0$
$\lambda_t = 0.03000$ $\quad\quad$ $\mu_r = 0.02500$ $\quad\quad$ $\mu_c = 0.0025$
$\lambda_b = 0.02000$ $\quad\quad$ $\lambda_r = 0.00210$ $\quad\quad$ $\lambda_g = 0.0$
$\mu_s = 0.03500$ $\quad\quad$ $\mu_b = 0.02400$ $\quad\quad$ $\mu_g = 0.0030$

Probability of failure in State 2 $(P_{2-3} + P_{2-4} + P_{2-5})$ for
Lead time (2 hours) $\ = 0.0041595$
Lead time (4 hours) $\ = 0.0080157$
Lead time (6 hours) $\ = 0.0116029$
Lead time (8 hours) $\ = 0.0148479$
Lead time (12 hours) $= 0.0210208$

The study results are shown in Table 5.8. It can be seen that rapid start units have a greater effect on the risk levels than the units on hot reserve. Table 5.8 shows the risk levels in the hypothetical system as a function of the hot reserve capacity at different levels of rapid start capability.

5.3.4 Practical Application

The concepts illustrated previously have been applied to the Saskatchewan Power Corporation (SPC) System. The on-line operating capacity was taken to be 1128.9 MW. The system was assumed to have gas turbine and gas engine units at Estevan and Success with a total capacity of 25 MW. Three units at the A. L. Cole Generating Station with a total capacity of 46.3 MW were assumed to be on hot reserve. The risk levels were then computed including the effect of rapid start generation and the units on hot reserve. The start-up time of the hot reserve units was considered to be 1 hour. Two values of lead time, 5 minutes and 10 minutes, were considered for the rapid start generation. The risk levels for various reserve margins are shown in Figure 5.15 for rapid start generation lead times of 5 minutes and 10 minutes.

As noted previously, the system load cannot be considered to be a single value. The area risk curves are modified by both load and operating capacity changes. The role of rapid start units, conventional hydro and pumped storage hydro units, and conventional thermal generation can be included in a spinning-reserve analysis. This approach should prove useful in practical utility studies where the lead time associated with additional or standby capacity cannot be considered as a single constant value. The economics associated with a given generation composition can be readily determined and the benefits associated with

Generation-System Operation

Table 5.8 Variation in Risk Level for Different Amounts of Generation on Hot Reserve

Operating capacity = 2100 MW
Load = 1900 MW
Risk without any rapid start generation = 0.00239083

Hot Reserve MW	Risk in Period 1	Risk in Period 2	Risk in Period 3	Total Risk
(a) Rapid Start Generation Available = 30 MW				
0	0.00008399	0.00004199	0.00054262	0.00066860
50	0.00008399	0.00004199	0.00019673	0.00032271
100	0.00008399	0.00004199	0.00005107	0.00017705
150	0.00008399	0.00004199	0.00000714	0.00013312
200	0.00008399	0.00004199	0.00000150	0.00012748
250	0.00008399	0.00004199	0.00000043	0.00012641
300	0.00008399	0.00004199	0.00000005	0.00012603
(b) Rapid Start Generation Available = 60 MW				
0	0.00008399	0.00001389	0.00021113	0.00030901
50	0.00008399	0.00001389	0.00005543	0.00015331
100	0.00008399	0.00001389	0.00000844	0.00010632
150	0.00008399	0.00001389	0.00000168	0.00009956
200	0.00008399	0.00001389	0.00000043	0.00009831
250	0.00008399	0.00001389	0.00000004	0.00009792
300	0.00008399	0.00001389	0	0.00009788
(c) Rapid Start Generation Available = 90 MW				
0	0.00008399	0.00001316	0.00019249	0.00028964
50	0.00008399	0.00001316	0.00003867	0.00013582
100	0.00008399	0.00001316	0.00000319	0.00010034
150	0.00008399	0.00001316	0.00000126	0.00009841
200	0.00008399	0.00001316	0.00000033	0.00009748
250	0.00008399	0.00001316	0	0.00009715
300	0.00008399	0.00001316	0	0.00009715
(d) Rapid Start Generation Available = 150 MW				
0	0.00008399	0.00000245	0.00003545	0.00012189
50	0.00008399	0.00000245	0.00000292	0.00008936
100	0.00008399	0.00000245	0.00000124	0.00008768
150	0.00008399	0.00000245	0.00000032	0.00008676
200	0.00008399	0.00000245	0	0.00008644
250	0.00008399	0.00000245	0	0.00008644
300	0.00008399	0.00000245	0	0.00008644

The Effect of Peaking Equipment on the Operating Reserve 157

Hot Reserve MW	Risk in Period 1	Risk in Period 2	Risk in Period 3	Total Risk
(e) Rapid Start Generation Available = 300 MW				
0	0.00008399	0	0.00000030	0.00008429
50	0.00008399	0	0	0.00008399
100	0.00008399	0	0	0.00008399
150	0.00008399	0	0	0.00008399
200	0.00008399	0	0	0.00008399
250	0.00008399	0	0	0.00008399
300	0.00008399	0	0	0.00008399

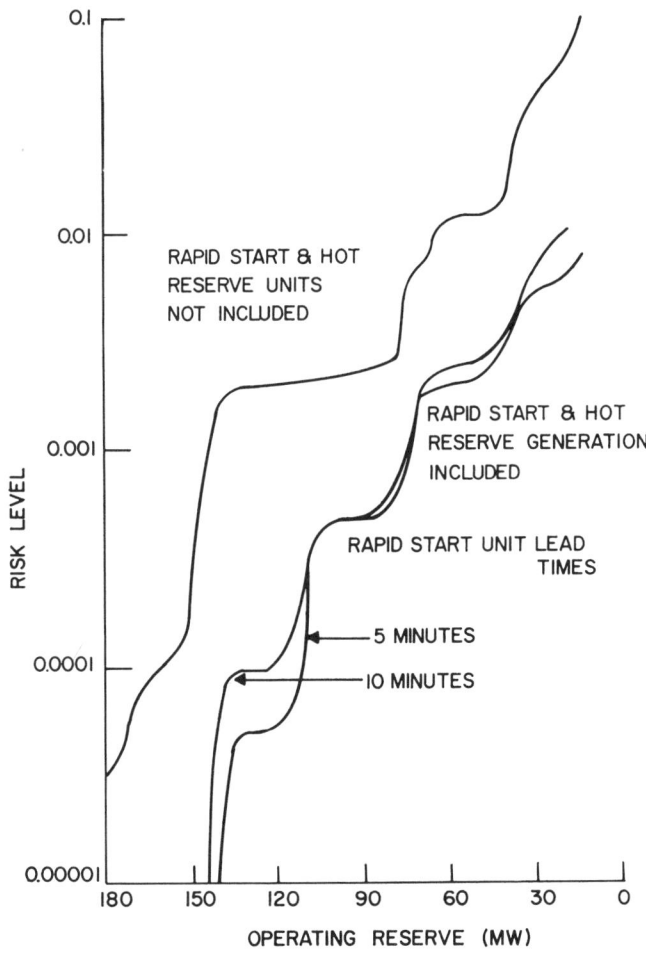

Figure 5.15 Effect of rapid start and hot reserve units on the SPC risk levels.

rapid start equipment evaluated using a predetermined risk index for the system. The effects of unit deratings and load forecast uncertainties can be included in the analysis using the previously illustrated techniques.

5.4 Interconnected System Spinning-Reserve Requirements

There has been a considerable amount of work done in the area of static capacity reliability evaluation in interconnected systems. The first important paper in this subject[10] utilized a two-dimensional probability array upon which the tie-line constraints were imposed. The basic generating-unit statistics in static and spinning-reserve studies are fundamentally different, and in the latter case the problem is a question of whether the capacity immediately available to the system will be capable of satisfying the load in the event of possible capacity and load changes.[11] The possible assistance under these conditions from an interconnected system cannot therefore be neglected in determining the risk of just carrying or failing to carry the system load. In a static-capacity interconnected-system study, the tie line is normally rated at its nominal value. In a spinning study, the tie-line rating or intersystem transfer capability will depend upon the immediate and known conditions between each system, that is, the existing transmission configuration, the scheduled generation pattern, and so forth. These could be quite variable even over the course of a day.

The approach used to determine the benefits of interconnection has been designated as the "Capacity Assistance Method" and is quite fast and efficient in regard to computer storage.[3,12]

5.4.1 Two-System Probability Array Approach

Cook and his co-workers, in their paper,[10] proposed the use of a two-dimensional capacity outage array to find the probability of simultaneous outages in the two systems. If the outage in the first system is less than the operating reserve, it can assist the second system to the extent of the effective assistance or the tie-line capability, whichever is less. In the spinning-reserve sense, the operating reserve is the additional capacity synchronized to the bus and available to pick up load. The effective assistance in this case is the operating reserve minus the capacity on outage. If the outage in the second system is less than the effective as-

Interconnected System Spinning-Reserve Requirements 159

sistance, the risk for that particular simultaneous event is zero. If, however, the effective assistance fails to satisfy the given capacity outage, the probability of occurrence of the simultaneous outages in the two systems will contribute to the risk level. If the operating reserve in the second system is greater than the capacity on outage, the event does not contribute to the risk level. The probabilities of all simultaneous conditions that lead to capacity deficiencies are computed and added. This sum is the risk level for the particular operating capacities and the corresponding loads in the two systems. This method becomes a little unwieldy for the calculation of system risk levels over a long period of time for varying operating capacities in the two systems.

5.4.2 Capacity Assistance Probability Method

The individual system risk levels can be obtained using the capacity assistance probability method. A probability table for the assisting system is first obtained. This assistance table is a two-dimensional array representing the amount of assistance and the corresponding probability of assistance. This table can be converted into a capacity-outage probability table with total capacity equal to the maximum assistance available by replacing each level of capacity available with a capacity equal to the maximum assistance minus the actual capacity assistance. The effective assistance as modified by the tie-line capability is essentially equivalent to a single unit with many derated states. This equivalent single unit with a capacity equal to the maximum possible assistance can be added to the capacity-outage probability table of the assisted system. The resulting capacity-outage probability table can be used to determine the risk levels in the assisted system.

The basic algorithm for this approach is as follows: If System A has operating capacity $= C_A$ MW,
$$\text{load} = L_A \text{ MW},$$
operating reserve $S_A = C_A - L_A$ MW,
and
tie-line capacity $= TL$ MW,
then $P(C)$ is the cumulative outage probability and $T(C)$ is the individual outage probability of C MW. See Table 5.9.

Two distinct cases arise.

Case a: $S_A \leq TL$.

Generation-System Operation

Table 5.9 Capacity-Outage Probability Table for System A

Capacity Out (MW)	Cumulative Probability	Individual Probability
$C_0 = 0$	$P(C_0)$	$T(C_0)$
C_1	$P(C_1)$	$T(C_1)$
C_2	$P(C_2)$	$T(C_2)$
⋮	⋮	⋮
C_n	$P(C_n)$	$T(C_n)$

Table 5.10 Equivalent Assistance Unit of System A for Case a

Capacity Out (MW)	Probability
C_0	$T(C_0)$
C_1	$T(C_1)$
C_2	$T(C_2)$
⋮	⋮
C_k	$P(C_k)$

The equivalent assistance unit of System A for Case a is given in Table 5.10.

Here $C_k \leq$ TL.

Case b: $S_A >$ TL.

Let $C_L = S_A -$ TL.

The equivalent assistance unit of System A for Case b is given in Table 5.11.

In Table 5.11, m indicates the position in the capacity outage probability table of the assisting system at which the capacity outage is C_L MW; k is the number of steps in the equivalent assistance unit of System A.

Table 5.11 Equivalent Assistance Unit of System A for Case b

Capacity Out (MW)	Probability
$C_m - C_L = 0$	$1 - P(C_{m+1})$ where $C_m = C_L$
$C_{m+1} - C_L$	$T(C_{m+1})$
$C_{m+2} - C_L$	$T(C_{m+2})$
$\vdots$	$\vdots$
$C_{m+k} - C_L = \text{TL}$	$P(C_{m+k})$

Table 5.12 Capacity-Outage Probability Table for System A

Capacity Out (MW)	Cumulative Probability	Individual Probability
0.0	1.0000	0.9642
10.0	0.0358	0.0062
15.0	0.0296	0.0008
20.0	0.0288	0.0012
30.0	0.0276	0.0170
40.0	0.0106	0.0018
60.0	0.0088	0.0001
70.0	0.0087	0.0005
80.0	0.0082	0.0072
90.0	0.0010	0.0005
100.0	0.0005	0.0005

This algorithm may be used to develop the equivalent assistance unit of the assisting system.

The approach can be easily illustrated using a simple numerical example.

Consider the capacity outage probability table of System A with the following characteristics (see Table 5.12):
operating capacity = 200 MW,
tie-line capability = 50 MW,

and
$$\text{peak load} = 160 \text{ MW}.$$
Clearly,
$$\text{maximum assistance} = \text{Min } \{200-160, 50\}$$
$$= 40 \text{ MW}.$$

See the various assistance probabilities in Table 5.13 for this case. The equivalent multiderated state assistance unit probabilities for this case are given in Table 5.14.

Table 5.13 Assistance Probability Table for System A

Assistance from System A	Probability
40.0 − 0.0 = 40.0	0.9642
40.0 − 10.0 = 30.0	0.0062
40.0 − 15.0 = 25.0	0.0008
40.0 − 20.0 = 20.0	0.0012
40.0 − 30.0 = 10.0	0.0170
40.0 − 40.0 = 0.0	0.0106

Table 5.14 Equivalent Multiderated State Assistance Unit

Capacity Out (MW)	Probability
0.0	0.9642
10.0	0.0062
15.0	0.0008
20.0	0.0012
30.0	0.0170
40.0	0.0106

Table 5.15 Equivalent Multiderated State Assistance Unit for 140-MW Peak Load

Capacity Out (MW)	Probability
0.0	0.9704
5.0	0.0008
10.0	0.0012
20.0	0.0170
30.0	0.0018
50.0	0.0088

Consider the case when

peak load = 140 MW

and

tie-line capability = 50 MW.

The equivalent multiderated state assistance unit probabilities for this case are given in Table 5.15.

5.4.3 Fundamentals of Interconnected System Assistance

There are two basic factors determining the extent of the interconnection assistance from one system to the other. They are the operating reserve in the assisting system and the tie-line transfer capability.

The assumption has been made that any negative capacity contingency in the assisting system does not affect the assisted system. Positive capacity margins in either system do, however, increase the reliability of the other system. When the tie-line capacity increases beyond the operating reserve of the assisting system, no further gain in reliability is obtained by increasing the tie-line capability. This limiting tie-line capacity is called the infinite tie capability. If the tie-line capability is reduced beyond the operating reserve, it limits the maximum assistance to the maximum transfer capability. This is visually seen from the capacity assistance probability table, which does not change after

Generation-System Operation

$TL > S_A$. The operating reserve in the assisting system determines the extent of the assistance up to the tie-line transfer capability. After which, if the operating reserve is further increased, the reliability of the assistance increases; that is, the probabilities of higher assistance are increased while those of lesser assistance are decreased.

The maximum possible reliability, that is, the minimum possible risk level in the assisted system, is basically determined by the assisted system itself. If the assisted system has a tie line whose maximum capacity is T_M MW, then the lowest risk ever possible for that system with infinite interconnected system assistance at a load of L_A MW will be $P(M_{min})$, where

$P(M_{min})$ is the cumulative outage probability of M_{min} MW

and

$$M_{min} = C_A - L_A + T_M.$$

Here

C_A = operating capacity of the assisted system

and

L_A = load on the assisted system.

5.4.4 Inclusion of the Tie-Line Forced Outage Rate

The discussion presented earlier has assumed a perfect tie line that is assumed to be always available. In actual practice the tie line may be forced out of service due to a variety of reasons. The tie-line availability can be readily incorporated into the equivalent assistance unit of the system.

Let the tie-line forced outage probability $= p$ and the tie-line availability $q = 1 - p$.

When the tie line is out, no assistance is possible and assistance to the other system is available only when the tie line is available.

Interconnected System Spinning-Reserve Requirements

The equivalent assistance unit can be modified by the tie-line availability.

Case a: $S_A \leq TL$.

The equivalent assistance unit of System A for this case is shown in Table 5.16.

Case b: $S_A > TL$.

Let $C_L = S_A - TL$.

The equivalent assistance unit of System A for this case is given in Table 5.17.

Table 5.16 The Equivalent Assistance Unit of System A for Case a

Capacity Out (MW)	Probability
$C_0 = 0$	$T(C_0) \cdot q$
C_1	$T(C_1) \cdot q$
C_2	$T(C_2) \cdot q$
$\vdots$	$\vdots$
$C_k = S_A$	$P(C_k) \cdot q + p$

Table 5.17 The Equivalent Assistance Unit of System A for Case b

Capacity Out (MW)	Probability
$C_m - C_L = 0$	$[1 - P(C_{m+1})] \cdot q$
$C_{m+1} - C_L$	$T(C_{m+1}) \cdot q$
$C_{m+2} - C_L$	$T(C_{m+2}) \cdot q$
$\vdots$	$\vdots$
$C_{m+k} - C_L = TL$	$P(C_{m+k}) \cdot q + p$

In Table 5.17, m indicates the position in the capacity outage probability table of the assisting system at which the capacity outage is C_L MW.

This modified table will include the effect of tie line outages.

5.4.5 Practical Application

In order to illustrate the practical application of interconnected system operating reserve evaluation, the Saskatchewan Power Corporation (SPC) system interconnected to the Manitoba Hydro (MH) system was studied.[11]

The operating capacity level of the SPC system was considered as 1130 MW, while the operating capacity level of Manitoba Hydro was considered to be 1361.4 MW. The most important factors affecting the interconnection benefits in the order of their importance are

1. operating reserve in Manitoba Hydro,
2. tie-line capability,

and

3. tie-line forced outage rate.

In Study 1, the tie-line capability was fixed at 135 MW, while the load in the Manitoba Hydro system was changed from 1201.4 to 1326.4 MW in steps of 25 MW. The variation of the SPC risk level versus peak load is given in Figure 5.16. It can be seen that the operating reserve in the Manitoba Hydro system significantly affects the risk in SPC. As the load increases in the Manitoba Hydro system, MH, the risk levels become higher in the SPC system.

In Study 2, the operating reserve in the Manitoba Hydro system was fixed at 160 MW, while the transfer capability of the tie line was varied from 75 to 300 MW. When the tie-line transfer capability was increased from 75 to 100 MW and 100 to 135 MW, significant decreases in the risk level are visible. As previously noted, the tie-line transfer capability may vary over the course of a day as the conditions within the two systems vary. The gain in the risk level is relatively small when the transfer capability is further increased from 135 to 175 MW. As explained earlier, the infinite tie capacity in this case is 160 MW. No gain at all in the risk level was obtained after the transfer capability was increased from 175 MW to 300 MW. These curves are shown in Figure 5.17.

Reference 11 shows the variation in the SPC risk levels when the

Interconnected System Spinning-Reserve Requirements

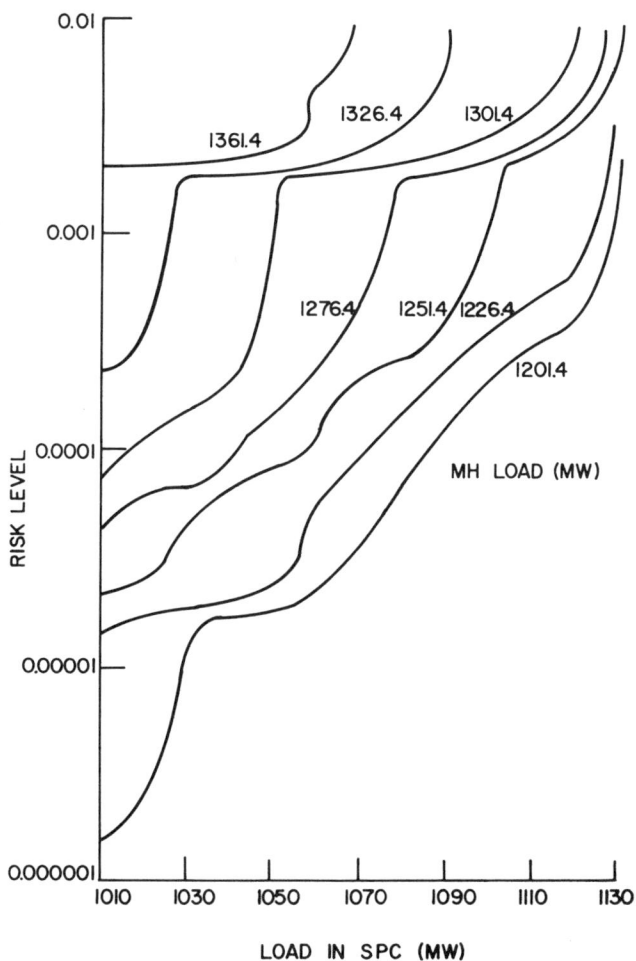

Figure 5.16 Effect on the SPC risk levels of variation in the operating reserve in the MH system, tie-line transfer capability 135 MW.

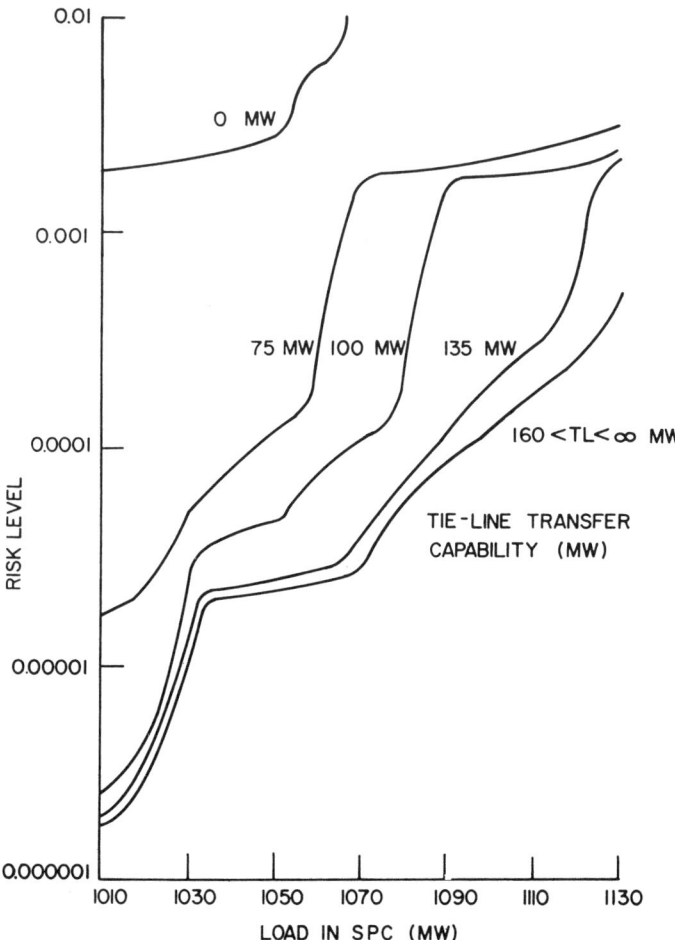

Figure 5.17 Effect on the SPC risk levels of variation in tie-line transfer capability, operating reserve in the MH system 160 MW.

tie-line forced outage probability is varied. As expected, the risk level increases as the probability of the tie-line availability decreases.

The effect of load forecast uncertainty in either or both systems can be included in the analysis in a similar manner to that employed in single-system studies.[11] The generating unit models in each system should include the possible derating levels and any rapid start equipment available.

Methods have been presented for calculation of the spinning-reserve interconnection benefits in two interconnected systems. The concept of a single system being equivalent to one gigantic unit having a large number of derated states can be utilized to incorporate the practical limitations of tie-line constraints and the generating-unit derated states. The methods presented can be used to obtain an economic appraisal of the benefits associated with ensuring various tie-capacity levels. The nature of the agreement between the systems influences the interconnection benefits, and certain agreements may favor only one particular system. With some modifications the effect of these agreements can be incorporated.

References

1. L. T. Anstine, R. E. Burke, J. E. Casey, R. Holgate, R. S. John, and H. G. Stewart, "Application of Probability Methods to the Determination of Spinning Reserve Requirements for the Pennsylvania–New Jersey–Maryland Interconnection," *IEEE, Transactions on Power Apparatus and Systems*, vol. 82, 1963, pp. 726–735.

2. R. Billinton and M. P. Musick, "Spinning Reserve Criteria in a Hydro Thermal System by the Application of Probability Mathematics," *Journal of the Engineering Institute of Canada*, vol. 48, October 1965, pp. 40–45.

3. R. Billinton, *Power System Reliability Evaluation*. New York: Gordon and Breach Science Publishers, 1970.

4. B. E. Biggerstaff and T. M. Jackson, "The Markov Process as a Means of Determining Generating Unit State Probabilities for use in Spinning Reserve Applications," *IEEE, Transactions on Power Apparatus and Systems*, vol. 88, 1969, pp. 428–428.

5. R. Billinton and A. V. Jain, "Unit Derating Levels in Spinning Reserve Studies," IEEE Paper No. 71 TP 120-PWR, 1971 Winter Power Meeting.

6. A. D. Patton, "Short Term Reliability Calculation," *IEEE, Transactions on Power Apparatus and Systems*, vol. 89, 1970, pp. 509–513.

7. Texas A & M Electric Power Institute, Edison Electric Institute Project RP 90-6, "Methods of Bulk Power System Security Assessment Probability Approach," November 1970.

8. R. Billinton and A. V. Jain, "The Effect of Rapid Start and Hot Reserve Units in Spinning Reserve Studies," IEEE Paper No. 71 TP 506-PWR, 1971 Summer Power Meeting.

9. "A Four State Model for Estimation of Outage Risk for Units in Peaking Service," Report of IEEE Task Group on Models for Peaking Service Units, Application of Probability Methods Subcommittee, IEEE Paper No. 71 TP 90-PWR, 1971 Winter Power Meeting.

10. V. M. Cook, C. D. Galloway, M. J. Steinberg, and A. J. Wood, "Determination of Reserve Requirements of Two Interconnected Systems," *IEEE, Transactions on Power Apparatus and Systems*, vol. 82, 1963, pp. 18–33.

11. R. Billinton and A. V. Jain, "Interconnected System Spinning Reserve Requirements," IEEE Paper No. 71 TP 576-PWR, 1971 Summer Power Meeting.

12. R. Billinton, M. P. Bhavaraju, and P. Thompson, "Power System Interconnection Benefits," *Transactions of the Canadian Electrical Association*, 1969.

INDEX

Adler, H. A., 4, 46, 62
Area risk curves, 141
Arnoff, E. L., 61
Availability, 1, 11
  capacity, 47, 56
Average cycle time, 12, 47, 49, 52
Average downtime, 12
Average uptime, 12

Baldwin, C. J., 62
Benner, P. E., 2
Biggerstaff, B. E., 153
Bulk supply systems, 4, 101
Bus configuration, ring, 35
Bus failure frequency, 113
Bus failure probability, 113

Calabrese, G., 4
Capacity, 47
  adequacy, 104
  availability, 47
  outage, 47
  reserve margin, 67
Capacity Assistance Method, 158
Capacity states
  cumulative, 53, 55
  derated, 131
  exact, 53
  identical, 55
  merged, 56
Chambers, J. C., 61
Circuit breaker, 9
  failure modes, 9
Component (device) arrangement
  parallel, 16
    frequency of failure, 17, 18
    repair (down) time, 17
    unavailability, 17
    uptime, 17
  series, 12–15
    availability, 13
    frequency of failure, 13, 14
    repair time, 14, 15
Composite data requirements, 114
Composite systems, 104

Conditional probability, 109
Cook, V. M., 5, 100
Cumulative frequency curves, 65
Cumulative states, 53
Cycle, 11
  average time of, 12, 47, 49, 52
  failure-repair, 11

Dale, K. M., 62
Dean, S. M., 1
Derated state, 131
  forced, 139
Duration
  expected or average, 11
  of capacity shortage, 81, 84
  to failure, 11
  to repair, 11

Edison Electric Institute, 66
Egly, D. T., 41
Einhorn, S. J., 61
Environmental effects, 24
Eppler, E. P., 31
Esser, W. F., 41

Failure
  bunching, 24
  common mode, 10, 29, 30, 31
  effect, 33
  mean time to (MTTF), 61
  mode, 9
  modes and effects analysis, 32, 37
  rate, 14
    normal weather, 24
    stormy weather, 24
    system, 101
Fault
  revealed, 9
  unrevealed, 9
Feeder
  overhead, 15, 20, 21
  underground, 15
Forced outage, 19, 24
Forced Outage Rate, 84, 141
  tie line, 164

Index

Fowler, P. H., 9
Frequency
  of cumulative capacity outage, 54
  of cumulative margin state, 77, 97
  event, 13
  fault, 12
  of merged state, 56
  of state encounter, 54

Generating capacity evaluation
  annual risk indices, 69, 78
  data requirements, 66
  expansion analysis, 79
  models
    multiarea, 101
    single area, 48, 101
    two area, 102
  planning studies, 46
Generation
  hydro forced outage rate, 19
  peaking, 145
Green, A. E., 9

Halperin, H., 4, 46, 62

Identical capacity states, 55
Independence, statistical, 12
Indices, performance
  availability, 1, 11, 101
  duration, 11, 101
  frequency, 12, 13, 101
Institute of Electrical and
    Electronics Engineers, Inc., 66
Interconnected systems
  limiting capacity, 105
  operating capacity, 163
  static capacity, 102
Interruption analysis, 21
Interval between outages, 3

Jackson, T. M., 153
Jacobs, I. M., 31

Lead time, 128
Load
  daily peak, 67, 68
    mean duration, 67, 68, 93
  model, 67, 69
  statistics, 90

Load-forecast uncertainty, 130
Loading, transformer, 27, 28
Load models, 67
Load statistics, 90
Loss-of-load probability, 3, 47, 52, 81, 87, 96
Lyman, W. J., 1

Maintenance, 23
  intervals, 69
Mallard, S. A., 41
Margin (reserve) states, 70, 73, 102
  availability, 72
  capacity reserve, 67
  cumulative, 74
  frequency of cumulative, 77
  of interconnected systems, 103
Markov chain, 67, 138
Markov process, 48, 133
Mean time-to-failure, 14, 61
Mean time-to-repair, 61
Models
  generating capacity, 47
  load, 67
  repairable system, 10
  two state, 12, 49

Outage
  event, 52
  forced, 19, 24
  generation, 47
Outage Replacement Rate (O.R.R.), 127
Overlapping outages, 20
Overload effects, 27

Parallel connected components, 16
Patton, A. D., 41
Peaking generating equipment, 141
Primary feeder, 20
Probability
  conditional, 109
  loss of load, 3, 47, 52, 81, 87, 96
  of system failure, 147
  transition, 135, 138
Process
  Markov, 48
  renewal, 10
  stochastic, 10

Index

Rate (transition)
  down, 54
  failure, 14, 49
  repair, 49, 149
  up, 54
Reliability, 1
  goal, 32, 36
  incremental benefits, 104
  prediction, 35
  procedure for substations, 31
Renewal process, 10
Repair, mean time to (MTTR), 61
Ring bus, 35

Sauter, D. M., 62
Security, 131
Series-connected components, 12
Smith, S. A., 1
State-space diagram, 133
State transition diagram, 51
Substation
  industrial, 41
  reliability procedures, 31
  ring bus, 35
Szendy, C., 5

Thomas, V. C., 41
Transformer overload risk, 27
Transition probability, 135, 138
Transition rate, determination, 149
Transmission, 101
  interconnection, 102
  planning, 124
  tie capacity, 104

Unavailability, 12, 49
  of service, 8
Uptime, 49